A Developer's Essential Guide to Docker Compose

Simplify the development and orchestration of multi-container applications

Emmanouil Gkatziouras

BIRMINGHAM—MUMBAI

A Developer's Essential Guide to Docker Compose

Group Product Manager: Rahul Nair

Publishing Product Manager: Niranjan Naikwadi

Senior Editor: Shazeen Iqbal

Content Development Editor: Romy Dias

Technical Editor: Arjun Varma

Copy Editor: Safis Editing

Project Coordinator: Ashwin Kharwa

Proofreader: Safis Editing

Indexer: Pratik Shirodkar

Production Designer: Joshua Misquitta

Marketing Coordinator: Nimisha Dua

First published: September 2022

Production reference: 1150922

Published by Packt Publishing Ltd.

Livery Place

35 Livery Street

Birmingham

B3 2PB, UK.

ISBN 978-1-80323-436-6

www.packt.com

To the amazing tech community in London. The vibrant tech community of London and its ecosystem helps me to be updated with the latest trends, to be motivated, and to interact with great engineers. Also, to my colleagues at Yapily.

Contributors

About the author

Emmanouil Gkatziouras started his software engineering journey when he joined a computer engineering and informatics department in Patras, Greece. He then worked as a software engineer for various companies. In 2015, he joined Oseven where he started working with cloud providers such as AWS and Azure, and container orchestration tools such as ECS and Kubernetes. He has fulfilled many roles, most recently as a cloud architect for the platform team.

He loves to give back to the developer community by contributing to open source projects such as InfluxDB, Spring Cloud GCP, and Alpakka and by blogging on various software topics. He is committed to continuous learning and is a holder of certifications such as CKA, CCDAK, PSM, CKAD, and PSO.

I want to thank myself, for giving me the time and support I've needed to write this book, on top of everyday responsibilities and priorities. I'd also like to thank my partner Viv for her patience while I was writing this book and the whole Packt editing team who assisted me: Romy Dias, Ashwin Dinesh Kharwa, and Niranjan Naikwadi.

About the reviewer

Werner Dijkerman is a freelance cloud, Kubernetes (certified), and DevOps engineer. He's currently focused on, and working with, cloud-native solutions and tools including AWS, Ansible, Kubernetes, and Terraform. He is also focused on Infrastructure as Code and monitoring the correct "thing" with tools such as Zabbix, Prometheus, and the ELK Stack, with a passion for automating everything and avoiding doing anything that resembles manual work.

Big thanks, hugs, and a shoutout to Ernst Vorsteveld!

Table of Contents

3

Network and Volumes Fundamentals   43

4

Executing Docker Compose Commands   63

Part 2: Daily Development with Docker Compose

5

Connecting Microservices 89

6

Monitoring Services with Prometheus 109

9

Creating Advanced CI/CD Tasks   161

Part 3: Deployment with Docker Compose

10

Deploying Docker Compose Using Remote Hosts   175

11

Deploying Docker Compose to AWS   189

12

Deploying Docker Compose to Azure 211

13

Migrating to Kubernetes Configuration Using Compose 223

Preface

The book explains the fundamentals of Docker Compose and its usage. You will discover the usage of Docker components under Compose along with Compose commands, their purpose, and their use cases. Further on, you will explore setting up databases for daily usage, leveraging Docker networking, and establishing communication between microservices. You will also run entire stacks locally on Compose, simulate production environments, and enhance CI/CD jobs using Docker Compose. Finally, you will learn about advanced topics such as Docker Compose on production deployments, provisioning infrastructure on public clouds such as AWS and Azure, and also pave the way for a migration to the Kubernetes orchestration engine.

Who this book is for

This book is for software engineers, developer advocates, and DevOps engineers looking to set up multi-container Docker applications using Compose without the need to set up a Docker orchestration engine and the expertise required. It is also for team leads looking to increase the productivity of an organization's software teams by streamlining the provisioning of complex development environments locally using Docker Compose.

What this book covers

Chapter 1, *Introduction to Docker Compose*, provides an overview of how Compose works and its various usages. There will be a brief explanation of the Docker Compose file format and a Compose example will be run.

Chapter 2, *Running the First Application Using Compose*, shows you how to create a simple Golang application that interacts with a Redis database. At the end of the chapter, you will have managed to run a multi-container application through Compose.

Chapter 3, *Network and Volumes Fundamentals*, dives into the fundamentals of Docker volumes and networks. At the end of the chapter, you will have defined and used a network for the existing application.

Chapter 4, *Executing Docker Compose Commands*, takes you through the Compose commands, their purpose, and the use cases.

Chapter 5, *Connecting Microservices*, explores creating new microservices. At the end of the chapter, you should have developed new microservices within the same network and established connectivity between them.

Chapter 6, Monitoring Services with Prometheus, covers adding monitoring to the services backed by the monitoring solution Prometheus.

Chapter 7, Combining Compose Files, looks at modularizing the Compose file and splitting it into multiple parts.

Chapter 8, Simulating Production Locally, provides an overview of complex Compose configurations with the goal of simulating production partially or fully in a local environment.

Chapter 9, Creating Advanced CI/CD Tasks, shows you how to create more advanced CI/CD tasks by simulating cases using Compose.

Chapter 10, Deploying Docker Compose Using Remote Hosts, covers deploying to remote hosts using Compose.

Chapter 11, Deploying Docker Compose to AWS, covers utilizing the knowledge acquired on Compose to achieve a deployment on AWS using ECS.

Chapter 12, Deploying Docker Compose to Azure, focuses on another popular cloud provider, Azure. At the end of the chapter, you should achieve a deployment on Azure ACI.

Chapter 13, Migrating to Kubernetes Configuration Using Compose, shows you how to translate the Compose files to a Kubernetes Deployment.

To get the most out of this book

You are expected to understand containerization and must possess fundamental Docker knowledge. Also, you should be comfortable with shell scripting. Ideally, a UNIX workstation would be the best option to progress through the book. Most of the code and commands presented should also be able to run on Windows machines.

Software/hardware covered in the book	Operating system requirements
Docker, Docker Compose, AWS, Azure	Windows, macOS, or Linux
Bash	Linux

If you are using the digital version of this book, we advise you to type the code yourself or access the code from the book's GitHub repository (a link is available in the next section). Doing so will help you avoid any potential errors related to the copying and pasting of code.

Download the example code files

You can download the example code files for this book from GitHub at https://github.com/ PacktPublishing/A-Developer-s-Essential-Guide-to-Docker-Compose. If there's an update to the code, it will be updated in the GitHub repository.

We also have other code bundles from our rich catalog of books and videos available at https://github.com/PacktPublishing/. Check them out!

Download the color images

We also provide a PDF file that has color images of the screenshots and diagrams used in this book. You can download it here: https://packt.link/kD3i4.

Conventions used

There are a number of text conventions used throughout this book.

Code in text: Indicates code words in text, database table names, folder names, filenames, file extensions, pathnames, dummy URLs, user input, and Twitter handles. Here is an example: "Mount the created nginx.conf configuration file as another file in your system."

A block of code is set as follows:

```
type Task struct {
  Id          string `json:"id"`
  Name        string `json:"name"`
  Description string `json:"description"`
  Timestamp   int64  `json:"timestamp"`
}
```

When we wish to draw your attention to a particular part of a code block, the relevant lines or items are set in bold:

```
services:
  redis:
    image: redis
    ports:
      - 6379:6379
```

Any command-line input or output is written as follows:

```
$ curl --location --request POST 'localhost:8080/task/'
$ cat /etc/nginx/nginx.conf
```

Bold: Indicates a new term, an important word, or words that you see onscreen. For instance, words in menus or dialog boxes appear in **bold**. Here is an example: "Select **System info** from the **Administration** panel."

> **Tips or Important Notes**
> Appear like this.

Get in touch

Feedback from our readers is always welcome.

General feedback: If you have questions about any aspect of this book, email us at customercare@ packtpub.com and mention the book title in the subject of your message.

Errata: Although we have taken every care to ensure the accuracy of our content, mistakes do happen. If you have found a mistake in this book, we would be grateful if you would report this to us. Please visit www.packtpub.com/support/errata and fill in the form.

Piracy: If you come across any illegal copies of our works in any form on the internet, we would be grateful if you would provide us with the location address or website name. Please contact us at copyright@packt.com with a link to the material.

If you are interested in becoming an author: If there is a topic that you have expertise in and you are interested in either writing or contributing to a book, please visit authors.packtpub.com.

Share your thoughts

Once you've read *A Developer's Essential Guide to Docker Compose*, we'd love to hear your thoughts! Scan the QR code below to go straight to the Amazon review page for this book and share your feedback.

https://packt.link/r/1803234369

Your review is important to us and the tech community and will help us make sure we're delivering excellent quality content.

Part 1:
Docker Compose 101

This part will introduce us to Docker Compose and how it works behind the scenes. We will familiarize ourselves with Compose by developing and deploying a set of applications using Compose. We will also find out how Docker concepts that we use daily (such as networking and volumes) map to Compose. Lastly, we will have an overview of the available Compose commands and familiarize ourselves with executing them.

The following chapters will be covered under this section:

- *Chapter 1, Introduction to Docker Compose*
- *Chapter 2, Running the First Application Using Compose*
- *Chapter 3, Network and Volumes Fundamentals*
- *Chapter 4, Executing Docker Compose Commands*

1

Introduction to Docker Compose

As **Docker** has rapidly become part of our daily developments and deployments, **Docker Compose** is a tool that you will encounter frequently. You have probably read about it, used it, or you might even have stumbled upon it while browsing the official Docker documentation.

As day-to-day development becomes more complex, it's common for an application to interact with more than one software component. Applications that grow in popularity will face the need to separate the workloads and facilitate scaling. The separation of logic, along with responsibilities to multiple software components, is imminent. Docker has been giving solutions for simplifying the containerization, management, and isolation of an application's workloads. Docker Compose can assist in the development of modern multi-container applications and their deployment.

Docker Compose is a simple and effective tool. Utilizing its features, it can help to tackle the challenges faced on multi-container applications and increase productivity in day-to-day development. Apart from its usage in the development life cycle, it can also be a viable option for production deployments. This bridges the gap between your initial local developments and actual production deployment. This capability can be utilized to achieve a smooth transition to orchestration engines such as Kubernetes.

This chapter will be an overview of Compose, how it works, and its common use cases. We will install Docker Compose and create our first Compose file to run a software component of our choice. By diving more into the Compose file format, we will also apply some extra configurations and use one of our local images.

In this chapter, the following topics will be covered:

- Introducing Docker Compose and its usage
- Installing Docker Compose

- Understanding how Docker Compose works
- Your first Docker Compose file
- Using your Docker image in Docker Compose

Technical requirements

The code for this book is hosted on GitHub at `https://github.com/PacktPublishing/A-Developer-s-Essential-Guide-to-Docker-Compose`. In case of an update to the code, it will be updated on GitHub.

Introducing Docker Compose and its usage

Docker Compose is a tool for defining and running multi-container Docker applications. The configuration is achieved using YAML files, and through the **Docker Compose CLI** utility, we can provision and perform operations on the containers managed by Docker Compose.

Here is a list of features that Compose offers:

- Complex multi-container applications on a single host
- The isolation of Docker workloads
- Bootstrapping and the distribution of complex applications
- Multiple environments
- The ability to preserve data on application change
- The ability to update application versions
- Environment composition
- Reusable configurations
- The simulation of complex production environments
- The deployment of production applications

In this book, we will dive into the preceding features extensively, evaluate how we can benefit from them, and incorporate them into our development process. In the next section, we will install Docker and Compose on our workstation using the operating system of our choice.

Installing Docker Compose

Both Docker Compose and the Compose CLI are built using **Go**. Compose can be run on the three major operating systems: Linux, Windows, and macOS. Since Compose is about managing multi-container Docker applications, the prerequisite is to have Docker installed.

Docker Desktop

On Mac and Windows, **Docker Desktop** is an installation option. Docker Desktop handles the complexity of setting up Docker on your local machine. It will create a Linux **Virtual Machine (VM)** on your host and facilitate container interactions with the OS such as access to the filesystem and networking. This one-click installation comes with the necessary tools such as the Docker CLI. One of the tools that is included is also Docker Compose. Therefore, installing Docker Desktop makes it sufficient to interact with Docker Engine using Compose on our workstation.

Installing Docker

To install the correct Docker distribution for the workstation of our choice, we will navigate to the corresponding section of the official Docker page:

- Docker Desktop for Mac: `https://docs.docker.com/desktop/mac/install/`

- Docker Desktop for Windows: `https://docs.docker.com/desktop/windows/install/`

- Docker Engine for Linux: `https://docs.docker.com/engine/install/`

On macOS

Apple provides workstations with two different types of processors: an Intel processor and an Apple processor. Docker has an installation option for both. Once the download is complete, by clicking on the installer, you can drag and drop the Docker application, as shown in the following screenshot:

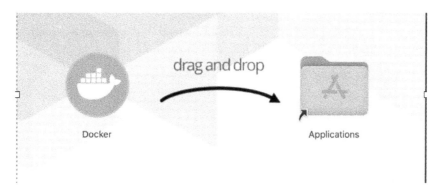

Figure 1.1 – Installing Docker on Mac

Once Docker has been installed, we can run a `hello world` command check:

```
$ docker run --rm hello-world
Unable to find image 'hello-world:latest' locally
```

```
latest: Pulling from library/hello-world
93288797bd35: Pull complete
Digest: sha256:97a379f4f88575512824f3b352bc03cd75e239179eea
0fecc38e597b2209f49a
Status: Downloaded newer image for hello-world:latest

Hello from Docker!
This message shows that your installation appears to be working
correctly.
..
```

Additionally, we have to check whether Compose has been installed:

```
$ docker compose version
Docker Compose version v2.2.3
```

Now, let's look at how to install Docker Desktop on Windows.

On Windows

Similar to Mac, Docker Desktop is installed seamlessly onto your **OS** ready to be used.

Once you download the EXE installation file and click on it, Docker will be installed along with its utilities. Once this is done, some extra configurations will need to be applied to enable virtualization for Windows.

Whether the backend that's being used is *WSL 2 backend* or *Hyper-V*, you have to set up your machine **BIOS** to enable virtualization, as shown in the following screenshot:

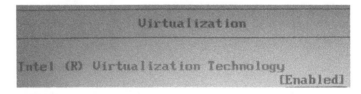

Figure 1.2 – Enabling virtualization on Windows via BIOS

Once you have logged in to Windows, you will need to enable the corresponding virtualization features.

For WSL 2, you should enable the **Virtual Machine Platform** feature and the **Windows Subsystem for Linux** feature:

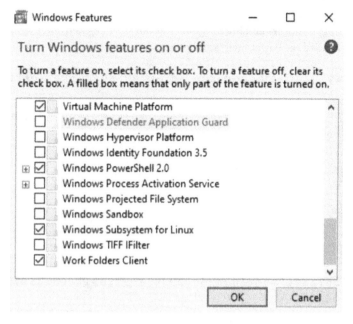

Figure 1.3 – Enabling virtualization for WSL 2

For Hyper-V you should enable **Hyper-V**:

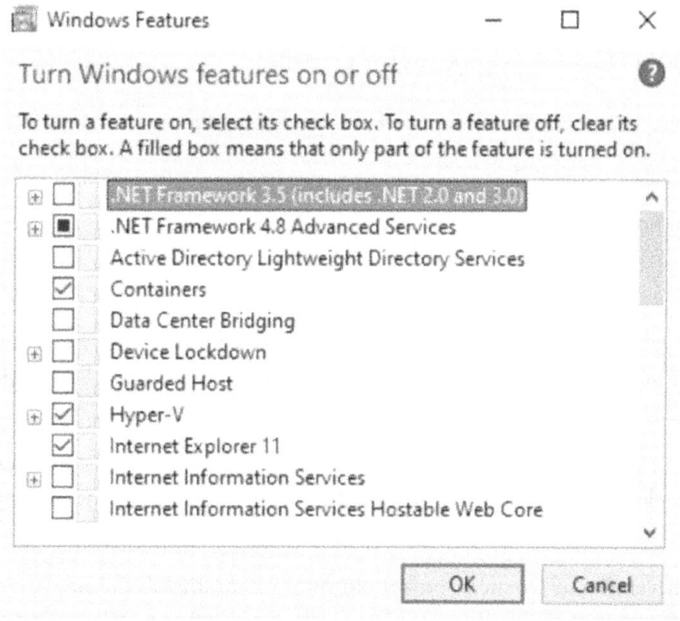

Figure 1.4 – Enabling virtualization for Hyper-V

Before you get started, make sure that your user account is added to the `docker-users` group. Once done, log out from Windows and log in again. You can start Docker, and then you can execute your first Docker command on PowerShell, as follows:

```
PS C:\Users\my-user> docker run -d -p 80:80 docker/getting-
started
Unable to find image 'docker/getting-started:latest' locally
latest: Pulling from docker/getting-started
59bf1c3509f3: Pull complete
8d6ba530f648: Pull complete
5288d7ad7a7f: Pull complete
39e51c61c033: Pull complete
ee6f71c6f4a8: Pull complete
f2303c6c8865: Pull complete
0645fddcff40: Pull complete
d05ee95f5d2f: Pull complete
Digest: sha256:aa945bdff163395d3293834697fa91fd4c725f47093ec499
f27bc032dc1bdd16
Status: Downloaded newer image for docker/getting-
started:latest
852371fcb34fddfe900bddc669af3a7aaab8743f8555fbb9952904bd2516a
e7a
PS C:\Users\my-user>
```

Let's also check whether Docker Compose has been installed:

```
PS C:\Users\my-user> docker compose version
Docker Compose version v2.2.3
```

Next, we will look at how to install Docker Desktop on Linux.

On Linux

At the time of writing, a Docker Desktop installation for Linux is not available, but it's on the roadmap, and it's just a matter of time before it'll be available for Linux. However, **Docker Engine** is sufficient in order to use Docker Compose.

The most common method of installation is to add the Docker repositories to your Linux workstation and then install Docker Community Edition using the corresponding package manager of the distribution used.

If you have an older version of Docker, you should remove and install the new `docker-ce` and `docker-ce-cli` versions. We will assume that this is the first Docker installation on the workstation we are currently using.

Since Red Hat-based Linux distributions are very popular for both workstations and production usage, we will install Docker on Fedora, which is a Red Hat-based distribution.

First, install the `dnf-plugins-core` package since it contains tools that can assist us with the management of the `dnf` repositories:

```
$ sudo dnf -y install dnf-plugins-core
```

Then, add the `docker-ce` repo to access the binaries provided by Docker:

```
$ sudo dnf config-manager --add-repo https://download.docker.
com/linux/fedora/docker-ce.repo
```

Now that the repo has been set up, we can add the packages:

```
$ sudo dnf install docker-ce docker-ce-cli containerd.io -y
```

Docker is a daemon that will run *as a service* to our machine. Therefore, the `systemctl` commands apply to Docker running *as a device*:

```
$ sudo systemctl start docker
```

Let's run a `hello-world` example:

```
$ sudo docker run hello-world


Hello from Docker!
This message shows that your installation appears to be working
correctly.
...
```

As you can see, we had to use `sudo` in almost every command. This can be fixed by having a group called `docker`, in which users will have the permission to interact with Docker Engine. On the installation of Docker Engine, this group will be created:

```
$ sudo groupadd docker
$ sudo usermod -aG docker $USER
$ docker run hello-world
```

Once installed, everything is set up to install Compose on Linux.

We will proceed with the installation link at https://docs.docker.com/compose/install/#install-compose-on-linux-systems:

```
$ sudo curl -L "https://github.com/docker/compose/releases/
download/1.29.2/docker-compose-$(uname -s)-$(uname -m)" -o /
usr/local/bin/docker-compose
$ sudo chmod +x /usr/local/bin/docker-compose
$ docker-compose —version
docker-compose version 1.29.2, build 5becea4c
```

Here, we can observe that this is an older version of Compose compared to the ones that we saw earlier. There isn't a standard way to install Compose V2 on Linux, for instance, by installing Docker Desktop on Mac and Windows. However, since it's feasible to install Compose V2 on Linux, we will proceed in doing so, allowing us to focus on Compose V2.

We will follow the guidelines from the official documentation at https://docs.docker.com/compose/cli-command/#install-on-linux:

```
$ mkdir -p ~/.docker/cli-plugins/
$ curl -SL https://github.com/docker/compose/releases/download/
v2.2.3/docker-compose-linux-x86_64 -o ~/.docker/cli-plugins/
docker-compose
$ chmod +x ~/.docker/cli-plugins/docker-compose

$ docker compose version
Docker Compose version v2.2.3
```

docker compose versus docker-compose

One observation to be made by navigating to the installation instructions for Linux is that a Python version of docker compose has been installed.

Also, this same version can be found on a Windows installation if you try to use the docker-compose command on Windows:

```
PS C:\Users\my-user> docker-compose-v1.exe version
docker-compose version 1.29.2, build 5becea4c
docker-py version: 5.0.0
CPython version: 3.9.0
OpenSSL version: OpenSSL 1.1.1g  21 Apr 2020
PS C:\Users\my-user>
```

The initial Docker Compose was built in Python; therefore, the installation instructions referenced the installation of `pip` packages.

Note that for new installations of Docker Desktop, the `docker-compose` command is an alias to `docker compose`.

The initial version of Compose's `docker-compose` is still supported and maintained. In the case of Compose applications built and run using `docker-compose`, there are supporting tools available such as **Compose Switch** (`https://docs.docker.com/compose/cli-command/#compose-switch`) for a smooth migration.

By installing Compose Switch, the old `docker-compose` command will be replaced by the `compose-switch` command.

Compose Switch will interpret the command that should have been passed to `docker-compose`. Then, it will translate it into a command that can be executed by Compose V2. Then, it will invoke Compose V2 using that command.

In this book, we shall focus on Compose V2 since it's part of `docker-cli`. This is the default on Docker Desktop, has the latest features, and comes with extra commands.

By now, you should have Docker and Docker Compose installed on your workstation and know how to execute some basic commands. You should also understand the previous Compose version and how you can transition to the latest version. Next, we're going to take a deeper dive into how Compose works and how it interacts with Docker Engine.

Understanding how Docker Compose works

Since we have Docker and Docker Compose installed onto our system, let's take some time and understand what Compose is and how it works behind the scenes.

On GitHub, we can find a project (`https://github.com/docker/compose`) where the Docker Compose source code is being hosted. By navigating to the source code, we can see and understand more about Compose, as follows:

- Compose integrates with the Docker CLI as a plugin.

- Compose interacts with Docker Engine through the API.

- Compose provides a CLI and its actions translate into Docker Engine API calls.

- Compose will read the Compose YAML file and generate resources accordingly.

- Compose provides a layer for converting `docker-compose` commands into CLI-compliant ones.

- Compose will interact with Docker objects and distinguish between them using labels.

The Docker CLI provides an API to create and load plugins. Once a plugin has been created and loaded on its invocation, the CLI command will be passed to it:

```
func pluginMain() {
    plugin.Run(func(dockerCli command.Cli) *cobra.Command {
        ...
        }
}

func main() {
    if commands.RunningAsStandalone() {
            os.Args = append([]string{"docker"}, compatibility.
Convert(os.Args[1:])...)
        }
    pluginMain()
}
```

The CLI is based on **Cobra** (https://github.com/spf13/cobra), which is a popular **Go** library for CLI applications.

Compose, being a plugin of the Docker CLI, will use a Docker Engine API client provided by the Docker CLI:

```
lazyInit.WithService(compose.NewComposeService(dockerCli.
Client(), dockerCli.ConfigFile()))
```

Each command passed to the Docker Compose plugin will lead to an interaction with the Docker Engine API on our host. For example, the internals of the ls command:

```
func (s *composeService) List(ctx context.Context, opts api.
ListOptions) ([]api.Stack, error) {
    list, err := s.apiClient.ContainerList(ctx, moby.
ContainerListOptions{
        Filters: filters.NewArgs(hasProjectLabelFilter()),
        All:      opts.All,
    })
    if err != nil {
        return nil, err
    }
```

```
        return containersToStacks(list)
}
```

We now have a good understanding of how Compose works and interacts with Docker Engine. You can also refer to the source code for more information. Next, we're going to run our first Docker Compose application.

Your first Docker Compose file

Imagine a scenario of wanting to run a static page on a server. For this task, an **NGINX** server is a good choice. We have a simple HTML file on the `static-site/index.html` path:

```
<!DOCTYPE html>
<html>
    <head>
        <title>Hello World</title>
    </head>
    <body>
        <p>Hi! This application should run on docker-compose</p>
    </body>
</html>
```

By using Docker, we will run an NGINX server using the official image found at `https://www.docker.com/blog/how-to-use-the-official-nginx-docker-image/`:

```
$ docker run --rm -p 8080:80 --name nginx-compose nginx
```

Let's break this down a little bit:

- Docker Engine will run a Docker NGINX image.

- The default port on the image is `80`, so we shall map it locally to `8080` to avoid using a privileged port.

- The name we assign will be constant in order to make interactions with the container easier.

- By using the `–rm` argument, we ensure that once we are done with our task and stop the container, the container will be deleted.

Our container is up and running. In a different Terminal session, we should access the default NGINX page:

```
$ curl 127.0.0.1:8080
<!DOCTYPE html>
<html>
<head>
<title>Welcome to nginx!</title>
<style>
html { color-scheme: light dark; }
body { width: 35em; margin: 0 auto;
font-family: Tahoma, Verdana, Arial, sans-serif; }
</style>
</head>
<body>
<h1>Welcome to nginx!</h1>
<p>If you see this page, the nginx web server is successfully
installed and
working. Further configuration is required.</p>

<p>For online documentation and support please refer to
<a href="http://nginx.org/">nginx.org</a>.<br/>
Commercial support is available at
<a href="http://nginx.com/">nginx.com</a>.</p>

<p><em>Thank you for using nginx.</em></p>
</body>
</html>
```

Since we have successfully run NGINX, we need to adapt our command in order to use the customized HTML page. A simple and fast way to do this is to mount the file at the path of a container. Let's exit the previous command using *Ctrl + C* and then refine the previous command:

```
docker run --rm -p 8080:80 --name nginx-compose -v $(pwd)/
static-site:/usr/share/nginx/html nginx
```

As expected, the page changes to the one we have specified:

```
$ curl localhost:8080/index.html
<!DOCTYPE html>
<html>
    <head>
        <title>Hello World</title>
    </head>
    <body>
        <p>Hi! This application should run on docker-compose</p>
    </body>
</html>
$
```

Now we have everything needed to migrate this application to Compose. We will create a Compose file for the default NGINX installation:

```
services:
  nginx:
    image: nginx
    ports:
      - 8080:80
```

Let's break down what we just did:

- The name of the service will be NGINX.
- The image is the same NGINX image.
- The ports are the same ports used previously.

The content shall be saved to a file named docker-compose.yaml.

Next, we will execute the Compose command on the Terminal:

```
$ docker compose up
[+] Running 2/0
  Network chapter1_default    Created
0.0s
  Container chapter1-nginx-1   Created
0.0s
```

```
Attaching to chapter1-nginx-1
chapter1-nginx-1  |  /docker-entrypoint.sh: /docker-
entrypoint.d/ is not empty, will attempt to perform
configuration
chapter1-nginx-1  |  /docker-entrypoint.sh: Looking for shell
scripts in /docker-entrypoint.d/
...
$
```

As expected, the result from the HTTP request is the same as the one that we experienced by just running the Docker container.

The naming of the file is important. We did execute the Compose command to spin up the Compose file, but we did not specify the file to be used. As it happens with `docker build` and Dockerfile, by running `docker compose` in a directory, Compose will search for a file named `docker-compose.yaml`. If the file exists, it'll be picked up as the default Compose file. Be aware that we are not limited to just one filename; we can use a different filename for our Compose applications. In the following chapters, there are cases where we can use a different name for the Compose files and run the application using the `-f` option.

Next, we shall mount the custom HTML page through the Compose configuration:

```
services:
  nginx:
    image: nginx
    ports:
      - 8080:80
    volumes:
      - ./static-site:/usr/share/nginx/html
```

As simple as our previous Docker command seemed to be, behind the scenes, it created a Docker volume pointing to a path of our filesystem and then it was attached to the container. The same applies to Compose. We specify a volume that points to our filesystem. Then, based on our location, it is mounted to a directory of the container:

```
$  curl localhost:8080/index.html
<!DOCTYPE html>
<html>
    <head>
        <title>Hello World</title>
    </head>
```

```
    <body>
        <p>Hi! This application should run on docker-compose</
p>
    </body>
</html>
```

As expected, the result is the same one with the result of the Docker example.

To review this section, we ran an NGINX instance using Docker CLI and made the transition to Compose by adding the corresponding YAML sections for the Docker command parameters that were used. Now, we are ready to move on to the next stage of this chapter's journey, where we'll build and run a Docker image on Docker Compose.

Using your Docker image on Docker Compose

By using Compose, we have achieved running the default NGINX image and changing the default HTML page that was displayed. Since we have started utilizing Compose, we will proceed with using and testing custom Docker images.

For our use case, we want to develop an NGINX image that prints logs in **JSON** format since it's feasible for tools such as **CloudWatch** (https://aws.amazon.com/cloudwatch/), **StackDriver** (https://cloud.google.com/products/operations), and **ELK Stack** (https://www.elastic.co/elastic-stack/) to persist data in JSON format and offer enhanced querying capabilities by having field conditions based on JSON elements.

The problem will require us to identify how NGINX defines the current logging format. Since we have a container already running through Compose, we will shell into the container and check the configuration:

```
$  docker ps
CONTAINER ID    IMAGE    COMMAND              CREATED
STATUS          PORTS                NAMES
dc0ca7ebe0cb    nginx    "/docker-entrypoint.…"   7 hours ago
Up 7 hours    0.0.0.0:8080->80/tcp    chapter1-nginx-1
$ docker exec -it chapter1-nginx-1 cat /etc/nginx/nginx.conf

user   nginx;
worker_processes   auto;

error_log  /var/log/nginx/error.log notice;
pid          /var/run/nginx.pid;
```

```
events {
    worker_connections  1024;
}
http {
    include         /etc/nginx/mime.types;
    default_type  application/octet-stream;

    log_format  main  '$remote_addr - $remote_user [$time_
local] "$request" '
                      '$status $body_bytes_sent "$http_referer" '
                      '"$http_user_agent" "$http_x_forwarded_for"';
```

By finding our running container using docker ps and issuing cat, through the container shell, we retrieved the current log_format from the instance by checking the /etc/nginx/nginx.conf file. We will change this format to JSON and build a custom Docker image preloaded with that format.

We will copy the file locally to apply the change:

```
$  docker cp chapter1-nginx-1:/etc/nginx/nginx.conf nginx.conf
```

By editing nginx.conf instead of log_format, we set the json format:

```
log_format  main escape=json '{"remote_addr":"$remote_
addr","remote_user":"$remote_user","time":"[$time_
local]","request":"$request",'
                      '"status":"$status","body_bytes_
sent":"$body_bytes_sent","http_referer":"$http_referer",'
                      '"http_user_agent":"$http_user_
agent","http_x_forwarded_for":"$http_x_forwarded_for"}';
```

Our file will look like this:

```
user  nginx;
worker_processes  auto;

error_log  /var/log/nginx/error.log notice;
pid        /var/run/nginx.pid;

events {
```

```
        worker_connections  1024;
}


http {
    include         /etc/nginx/mime.types;
    default_type  application/octet-stream;

    log_format  main  escape=json '{"remote_addr":"$remote_
addr","remote_user":"$remote_user","time":"[$time_
local]","request":"$request",'
                    '"status":"$status","body_bytes_
sent":"$body_bytes_sent","http_referer":"$http_referer",'
                    '"http_user_agent":"$http_user_
agent","http_x_forwarded_for":"$http_x_forwarded_for"}';

    access_log  /var/log/nginx/access.log  main;

    sendfile        on;
    #tcp_nopush      on;

    keepalive_timeout  65;

    #gzip  on;

    include /etc/nginx/conf.d/*.conf;
}
```

Now that we have the *config* file needed, we will create the base NGINX image that will use this configuration. The Dockerfile will be the following:

```
FROM nginx

COPY nginx.conf /etc/nginx/nginx.conf
```

Let's build the image:

```
$ docker build -t custom-nginx:0.1 .
```

Let's go ahead and use it with the recently created `docker-compose.yaml` file:

```
services:
  nginx:
    image: custom-nginx:0.1
    ports:
      - 8080:80
    volumes:
      - ./static-site:/usr/share/nginx/html


$ docker compose up

...

chapter1-nginx-1  | 2022/02/10 08:09:27 [notice] 1#1: start
worker process 33
chapter1-nginx-1  | {"remote_addr":"172.19.0.1","remote_
user":"","time":"[10/Feb/2022:08:09:33 +0000]","request":"GET
/ HTTP/1.1","status":"200","body_bytes_sent":"177","http_
referer":"","http_user_agent":"curl/7.77.0","http_x_forwarded_
for":""}

...
```

By now, Compose runs successfully on your application that also uses the custom Docker image. So far, Compose was sufficient to use a custom image and also include some modification at runtime such as mounting a file as well as doing port mapping. The results were the same as the ones we would expect if we run the application using Docker commands.

Summary

In this chapter, we were introduced to Docker Compose and some of its most notable features. We installed Compose on different operating systems and identified the differences between installations. Then, we identified the different Compose versions, Docker-Compose V1 and Docker Compose V2, along with the version to be used throughout this book. By checking on the Compose source code, we went a step further regarding how Compose works and interacts with the Docker CLI. Then, we ran a Docker application using the `docker-cli` command and created the equivalent of it on Compose. The next step was to customize the image we used in our first example and deploy it using Compose.

In the next chapter, we shall create an application that will run and interact with a Redis database using Compose.

2

Running the First Application Using Compose

In the previous chapter, we learned about Docker Compose and how we can benefit from its features. We learned how it works and interacts with the Docker engine and how it integrates with the Docker **command-line interface** (**CLI**). By installing **Compose** on the workstation of our choice, we were able to run some application examples.

Using Compose throughout an application's development can be a streamlined process that can have a significant role in increasing productivity. By the time you have completed this chapter, you will be able to package your application and run it on Compose. You'll be able to interact with the application and enhance its functionality by interacting with a database. Once the basic **multi-container application** is deployed using Compose, we shall dive deeper into more Compose functionalities such as health checks, labels, environment variables, and command override.

We will cover the following main topics in this chapter:

- Creating a core application
- Running Redis using Compose
- Shelling into a container managed by Compose
- Interacting with a Docker Compose service
- Packaging your application with Docker and Compose
- Running your multi-container application using Compose

Technical requirements

The code for the book is hosted on GitHub at `https://github.com/PacktPublishing/A-Developer-s-Essential-Guide-to-Docker-Compose`. If there is an update to the code, it will be updated on the GitHub repository.

Creating a core application

In this section, we will create an application using the **Go** programming language. Throughout the course of this book, this application will evolve and will help us display the features that Compose offers. Go is an open source programming language. Many software applications that we use nowadays are built using Go, including Docker, Kubernetes, and Compose. By programming using Go, we benefit from a robust library providing packages that we need such as `http server`, `testing utils`, and more.

Installing Go

Installing Go on your workstation is streamlined. By navigating to the download page on `https://go.dev/doc/install`, you can find installation packages for the major operating systems.

A REST API in Go using Gin

Building a **REST API** in Go can be done either by using the existing libraries or by using a framework. We will pick the framework option in order to keep our code base simple and keep our focus on Compose. The framework we will use is **Gin** (`https://gin-gonic.com/`), a popular Go framework that will assist us in routing, rendering JSON, and avoiding writing extra code should we use only the existing Go libraries.

Let's create our module with the following:

```
$ go mod init task_manager
```

We shall use an application template to get started:

```
$ curl https://raw.githubusercontent.com/gin-gonic/examples/
master/basic/main.go > main.go
```

Now, we are going to download Gin:

```
$ go get github.com/gin-gonic/gin
```

Following that, we will build the module using the `build` command of go:

```
$ go build
```

Now, you can run the application using go `run`:

```
$ go run main.go
```

Next, we shall test the application using `curl`:

```
$ curl http://localhost:8080/user/John
{"status":"no value","user":"John"}
```

We have successfully created our first REST API using Go. We have the tools to get started and develop our application. The next step is to build upon the template we just used and create our core application.

The application

The application will be a **Task Manager** available using a REST API interface. The application user will submit a task using a POST request. The user will also be able to retrieve the tasks using a GET request as well as delete them using a DELETE request. The endpoints we should cover for now are as follows:

- GET /task: Retrieve all tasks.

- POST /task: Add task.

- GET /task/{id}: Retrieve task by ID.

- DELETE /task/{id}: Delete task by ID.

With some adaptions to the main.go method, we will transform the class to the task server. The current implementation will keep the tasks in memory. The application can be in one file, as we can see at https://raw.githubusercontent.com/PacktPublishing/A-Developer-s-Essential-Guide-to-Docker-Compose/main/Chapter2/example-task-manager/main.go.

Our data structure will contain an ID, the name of the task, the description, and the timestamp:

```
type Task struct {
    Id          string `json:"id"`
    Name        string `json:"name"`
    Description string `json:"description"`
    Timestamp   int64  `json:"timestamp"`
}
```

Here, we can see the corresponding controllers for the actions mentioned:

```
// Get tasks
    r.GET("/task", func(c *gin.Context) {
        tasks := []Task{}
        for _, v := range taskMap {
            tasks = append(tasks, v)
```

```go
                }
            c.JSON(http.StatusOK, gin.H{"tasks": tasks})
        })
// Get task
        r.GET("/task/:id", func(c *gin.Context) {
            id := c.Params.ByName("id")
            task, ok := taskMap[id]
            if ok {
                    c.JSON(http.StatusOK, gin.H{"task": task})
            } else {
                    c.JSON(http.StatusNotFound, gin.H{"id": id,
"message": "not found"})
            }
        })
// Add task
        r.POST("/task", func(c *gin.Context) {
            var task Task
            if err := c.BindJSON(&task); err != nil {
                    c.JSON(http.StatusOK, gin.H{"task": task,
"created": false, "message": err.Error()})
            } else {
                    taskMap[task.Id] = task
                    c.JSON(http.StatusCreated, gin.H{"task": task,
"created": true, "message": "Task Created Successfully"})
            }
        })
// Remove task
        r.DELETE("/task/:id", func(c *gin.Context) {
            id := c.Params.ByName("id")
            delete(taskMap, id)
            c.JSON(http.StatusOK, gin.H{"id": id, "message":
"deleted"})
        })
```

Since our REST API is up and running, we will interact with it by issuing some calls using `curl`. To issue a *create task* request, enter the following command:

```
$ curl --location --request POST 'localhost:8080/task/' \
--header 'Content-Type: application/json' \
--data-raw '{
    "id": "8b171ce0-6f7b-4c22-aa6f-8b110c19f83a",
    "name": "A task",
    "description": "A task that need to be executed at the
timestamp specified",
    "timestamp": 1645275972000
}'
{"created":true,"message":"Task Created Successfully","tas
k":{"id":"8b171ce0-6f7b-4c22-aa6f-8b110c19f83a","name":"A
task","description":"A task that need to be executed at the
timestamp specified","timestamp":1645275972000}}
```

To issue a *get all tasks* request, enter the following command:

```
$ curl --location --request GET 'localhost:8080/task'
{"tasks":[{"id":"8b171ce0-6f7b-4c22-aa6f-
8b110c19f83a","name":"A task","description":"A
task that need to be executed at the timestamp
specified","timestamp":1645275972000}]}
```

To retrieve a specific task using a task ID, enter the following command:

```
$ curl --location --request GET 'localhost:8080/task/8b171ce0-
6f7b-4c22-aa6f-8b110c19f83a'
{"task":{"id":"8b171ce0-6f7b-4c22-aa6f-8b110c19f83a","name":"A
task","description":"A task that need to be executed at the
timestamp specified","timestamp":1645275972000}}
```

In order to keep our example requests simple and portable, we used `curl`. As Postman is a popular tool for interacting with REST APIs, a collection is available here: `https://github.com/PacktPublishing/A-Developer-s-Essential-Guide-to-Docker-Compose/blob/main/Chapter2/Task-Manager.postman_collection.json`.

So far, we've created a Task Manager application that provides a REST API. Then, by using `curl`, we tested the application's functionality. Onward, we will adapt the Task Manager application to use a Redis server for storage.

Running Redis using Compose

Our main application is set up. For simplicity purposes, instead of storing the tasks in a database, we used an *in-memory* map. This works well for prototyping; however, our data and tasks remain only in one process. In the case of spinning up two instances of our application, each instance will contain different tasks.

To tackle this, we will store the data in a database. This way, the data will be kept in one place and there will not be any differences in the tasks served by various instances. **Redis** will be our choice for storing the data.

Redis is a popular in-memory data structure storage. It's widely used as a cache, and all major cloud providers use it as a caching offering. Components such as **ElastiCache** in Amazon Web Services (https://aws.amazon.com/elasticache) or **Memorystore** in Google Cloud Platform (https://cloud.google.com/memorystore) have the option of using Redis. Also, Redis can be used for database purposes. Furthermore, it comes with *broker* and *streaming* capabilities. Throughout this book, Redis and its versatility will help us highlight the features of Docker Compose.

In order to run Redis, we will use Docker Compose. Therefore, let's create a Compose file with a Redis service:

```
services:
  redis:
    image: redis
    ports:
      - 6379:6379
```

The service name will be redis. The default port of Redis is 6379, so we shall bind it locally to port 6379. By running compose up, we should be able to interact with a redis database locally:

```
$ docker compose up
...
chapter2-redis-1  | 1:M 19 Feb 2022 14:18:20.302 # Server
initialized
chapter2-redis-1  | 1:M 19 Feb 2022 14:18:20.303 * Ready to
accept connections
```

A Redis database is up and running through Compose and is ready to accept connections. In the next section, we will shell into the Redis container, get familiar with it, and execute some commands.

Shelling into a container managed by Compose

Since we've been successful in running a Redis database using Compose, we will run some commands upon that Redis instance and get familiar with the database. As happens with many container-based distributions of databases, along with the actual database, the image can contain tools that help the user interact with the database for administrative or usage purposes.

The Redis CLI can be used to send commands to a Redis service. The Redis Docker image does contain the Redis CLI, so we should be able to use it with the running database.

Let's find our running Redis image:

```
$ docker ps --format "{{.Names}}"
chapter2-redis-1
$
```

Let's shell into this image:

```
$ docker exec -it chapter2-redis-1 bash
root@d189b089bcf6:/data#
```

We just shelled successfully into a container managed by Compose. Let's navigate and see what is already there:

```
$ root@d189b089bcf6:/data# ls
dump.rdb
root@d189b089bcf6:/data# ls /usr/local/bin
docker-entrypoint.sh  gosu  redis-benchmark  redis-check-aof
redis-check-rdb  redis-cli  redis-sentinel  redis-server
```

The default directory has a dump.rdb file. Also, by using ls, we can see the binaries such as redis-server, redis-sentinel, and redis-cli. Since the Redis CLI does exist on the image, we can execute some commands upon the running Redis server:

```
$ root@d189b089bcf6:/data# redis-cli
127.0.0.1:6379>
```

Redis will be the main storage for tasks, so we shall add a task ID using ZADD and then check the data stored:

```
127.0.0.1:6379> ZADD tasks 1645275972000 "8b171ce0-6f7b-4c22-
aa6f-8b110c19f83a"
(integer) 1
```

```
127.0.0.1:6379> ZRANGE tasks 0 -1 WITHSCORES
1) "8b171ce0-6f7b-4c22-aa6f-8b110c19f83a"
2) "1645275972000"
```

After running a Redis server using Compose successfully, we managed to shell into the Redis container, identified existing command-line tools, and used the `redis-cli` command to interact with the server. Next, we'll adapt our application in order to use the Redis server for storage purposes.

Interacting with a Docker Compose service

We ran Redis on Compose and we *shelled* to that instance in order to run some commands and add data. Obviously, interacting with that instance doesn't require us to shell on it. The instance has been configured to have the port `6379` bound to our local port `6379`.

For example, we should be able to interact with that instance by a `redis-cli` client that has `localhost` access. In the following, you can see another Redis Docker image accessing our Compose-managed Redis:

```
$ docker run --rm -it --entrypoint bash redis -c 'redis-cli -h
host.docker.internal -p 6379'
host.docker.internal:6379> ZRANGE tasks 0 -1 WITHSCORES
1) "8b171ce0-6f7b-4c22-aa6f-8b110c19f83a"
2) "1645275972000"
host.docker.internal:6379>
```

We can see that the entry that we added previously has been displayed on this terminal session. We are successfully interacting with the **Compose service** from the outside, so we shall proceed with adapting our application's code base in order to use Redis instead of an in-memory map. One structure to use would be a **sorted set**. A sorted set is like a set accompanied by a numerical value called a **score**. The elements on the set are sorted by the score. If they have the same score, then they are sorted based on the lexicographic order of the value. If we examine the Redis ZADD command that we ran previously, we can see that the score that we specified is actually the task's timestamp. So, the tasks on our services will be sorted using the timestamp.

Also, we want to retrieve tasks by `id`. **Hashes** are a good option since they are the perfect data type to represent objects. By using a **hash**, we can map and access the struct values in a key-value manner:

```
host.docker.internal:6379> HMSET task:8b171ce0-6f7b-4c22-
aa6f-8b110c19f83 Id 8b171ce0-6f7b-4c22-aa6f-8b110c19f83a Name
"A task" Description "A task that need to be executed at the
timestamp specified" Timestamp 1645275972000
OK
```

```
host.docker.internal:6379> HGETALL task:8b171ce0-6f7b-4c22-
aa6f-8b110c19f83
1) "Id"
2) "8b171ce0-6f7b-4c22-aa6f-8b110c19f83a"
3) "Name"
4) "A task"
5) "Description"
6) "A task that need to be executed at the timestamp specified"
7) "Timestamp"
8) "1645275972000"
host.docker.internal:6379>
```

Let's exit and terminate the container using *Ctrl + C*.

Now, we will adapt our application to use Redis for persistence. There are various client options, so we will choose the one that is currently more popular, which is the go-redis/redis client.

Let's import go-redis/redis on our project:

```
$ go get github.com/go-redis/redis/v8
```

After some changes, our code base should serve requests using Redis. We can see the Redis persistence operations on GitHub: https://github.com/PacktPublishing/A-Developer-s-Essential-Guide-to-Docker-Compose/blob/main/Chapter2/main.go#L91.

Also, the methods of the controller will be adapted accordingly, as we can see on GitHub: https://github.com/PacktPublishing/A-Developer-s-Essential-Guide-to-Docker-Compose/blob/main/Chapter2/main.go#L39.

For every task, there will be an equivalent hash. The prefix of the hash will be the string task. The hash will contain all the information submitted for a task. In order to have some sorting based on the date of the task, we will use a sorted set.

This brings us to the following Redis interactions per endpoint:

- GET /task: Retrieve task ID using the sorted set and fetch the hash for each task.

- POST /task: Add a hash named task:{id} for the task and insert the ID into the sorted set.

- GET /task/{id}: Retrieve the task hash using key task:{id}.

- DELETE /task/{id}: Remove the hash named task:{id} and remove the ID entry from the sorted set.

Let's run our application using the updated code base:

```
$ go run main.go
```

By running a Redis server using Compose, we've been able to migrate the existing core application to a Redis-backed solution. The code changes applied helped to replace the app-based storage with the appropriate Redis data structures for our use case. In the next section, we will package the application using Docker and run it using Compose.

Packaging your application with Docker and Compose

It's time that we moved on to packaging and deploying our application using Compose. In order to achieve this, the requisite is to create a Docker image for our application. Before we go in that direction, we need to adapt our code base so it can run in different environments without having to change the code and generate another image.

Enabling environment configuration

By examining the code base, we can see that certain configurations are subject to change. Redis configuration should be flexible since a Redis server, as long as it is accessible to our code base, can be located everywhere. For this reason, the Redis client will derive the configurations through environment variables. However, if there were no configurations provided, it should fall back to a default configuration.

The following utility function will assist us with this. In the case of no environment variable being provided, a default value will be used:

```go
func getIntEnv(key string, defaultvaule int) int {
    if value := os.Getenv(key); len(value) == 0 {
        return defaultvaule
    } else {
        if i, err := strconv.Atoi(value); err == nil {
            return i
        } else {
            return defaultvaule
        }
    }
}


func getStrEnv(key string, defaultValue string) string {
    if value := os.Getenv(key); len(value) == 0 {
```

```
            return defaultValue
        } else {
            return value
        }
    }
}
```

Onward, we will adapt the Redis client configuration to use **environment** variables:

```
var (
    client = redis.NewClient(&redis.Options{
            Addr:      getStrEnv("REDIS_HOST",
"localhost:6379"),
            Password: getStrEnv("REDIS_PASSWORD", ""),
            DB:        getIntEnv("REDIS_DB", 0),
    })
    taskMap = make(map[string]Task)
)
```

Finally, let's configure the host and port that our server will operate on:

```
func main() {
    r := setupRouter()
    r.Run(getStrEnv("TASK_MANAGER_HOST", ":8080"))
}
```

So far, we have adapted the existing code base to be configured through environment variables. This makes our solution portable in different environments. If no extra configuration is provided, our application can fall back to the default settings. We can now proceed to create a Docker image for the application.

Docker image creation

Thanks to our previous changes, our code base can now be containerized and run in different environments. This will be the Dockerfile for the application:

```
# syntax=docker/dockerfile:1
FROM golang:1.17-alpine
WORKDIR /app
COPY go.mod ./
COPY go.sum ./
```

```
RUN go mod download
COPY *.go ./
RUN go build -o /task_manager
EXPOSE 8080
CMD [ "/task_manager" ]
```

Let's examine the steps:

1. We specify the syntax to use when parsing the Dockerfile.

2. A golang alpine-based image is used.

3. /app is the directory the application will reside in and its dependencies.

4. mod and sum files are copied.

5. The command to download the dependencies is issued.

6. Our application files are copied.

7. The main application is built.

8. The port the application should run on is defined.

9. Our service is running.

Let's build our image:

```
$ docker build . -t task-manager:0.1
```

By creating a Dockerfile, we can now package our application in a Docker image and run it using the Docker CLI or Compose.

Running the image

Since our image is built, we can run our Docker image:

```
$ docker run --rm -p 8080:8080 --env REDIS_HOST=host.docker.
internal:6379 task-manager:0.1
```

The application will work as expected and as you can see, we specified it using the Redis host. Instead of using the Redis host, let's try to use the Redis container running on Compose. Since Redis is already running on Compose, it should be bound to the network created by Compose. We can find the network by listing the Docker networks present:

```
$ docker network ls
NETWORK ID       NAME              DRIVER      SCOPE
1392d8a87032     bridge            bridge      local
```

```
...
453bafcfd97a    chapter2_default    bridge    local
4a6d6a3e8475    host                host      local
...
```

We will mount the container to the `chapter2_default` network:

```
$ docker run --rm -p 8080:8080 --env REDIS_HOST=redis:6379
--network=chapter2_default task-manager:0.1
```

Take note that we changed the host. Because the container runs on the same network, the Redis instance can be accessed by using the service name, which will be resolved to the IP address inside the network.

Now that we have built the images instead of running them as standalone, we can add them to the Compose file:

```
services:
  task-manager:
    image: task-manager:0.1
    ports:
      - 8080:8080
    environment:
      - REDIS_HOST=redis:6379
  redis:
    image: redis
    ports:
      - 6379:6379
```

Let's break down how it works:

- Both services run on Compose.
- Since no network is specified, they use both the default network.
- By being on the same network, they can access each other using the service name.
- Any environment variables needed are set in the environment section.

We will recreate the application so that our changes will take effect:

```
$ docker compose up --force-recreate
```

Our multi-container application is now fully running on Compose, operating through the same network.

Build an image using Compose

By now, we have packaged our application fully using Compose. One of the things we spotted is that, throughout the chapter, our original application changed a lot. From a map-based Task Manager, it became an application backed by a Redis database with some configuration options. As expected, these will not be the last changes to the code base. The code base will get more changes in the future and, therefore, it's important to run our application on Compose in a more automated way.

Instead of building an image and changing to the new version using the image section of Compose, we should instruct Compose to build the image for us and run it. This is what you need to set up Compose to use the Dockerfile in order to build and use the image behind the scenes:

```
services:
  task-manager:
    build: .
    ports:
      - 8080:8080
    environment:
      - REDIS_HOST=redis:6379
  redis:
    image: redis
    ports:
      - 6379:6379
```

Since the image happens to have both the Dockerfile and `docker-compose.yaml` on the same directory, by using `build .`, we point to the same directory. In a scenario where the code base is on another image, we have the option to specify a path and run Compose with multiple directories and even different projects involved.

Building and defining the image name

So far, we got used to using the Task Manager through an *image name*. We will try to do the same thing on Compose by defining both `build` and `image`. By using `build` and `image`, the image will be built as expected and will then be tagged based on the name given using `image`:

```
services:
  task-manager:
    build: .
    image: task-manager:0.1
```

From building a Docker image using the Docker CLI manually, we executed the `build` process using Compose. The foundation for using Compose with our application has been established, so we can proceed with more advanced concepts of Compose.

Running your multi-container application using Compose

By having our multi-container application up and running on Compose, we can now jump on more specific concepts and identify how to get the most out of our application on local development.

Health check

Our code base has an extra endpoint that we did not mention previously, known as the **ping endpoint**. As seen on the code base, it's an endpoint replying with a constant message once invoked:

```
// Health Check
r.GET("/ping", func(c *gin.Context) {
    c.String(http.StatusOK, "pong")
})
```

This endpoint will be leveraged as a **health check**. As long as the application is running, it will always give back a response. Should there be something wrong with the application, the endpoint will not reply, and so will be marked *unhealthy*.

How health check works

By adding a health check using Compose, we instruct a command to be run on the container. Based on how the command exits and the time threshold, the container will be marked as *healthy* or *unhealthy*. The command will be executed inside the container. Using a non-existing command will result in a failed health check.

The command can be everything executable on the container. It can be an existing command on the container or even a customized script with a complex health check algorithm. What Compose will consider is the successful execution of the script within the right time limit.

Since our health check endpoint is implemented on `http server`, we need a command to run on it. **cURL** is a very good choice; therefore, we will adapt the Dockerfile and add it to the image:

```
# syntax=docker/dockerfile:1
FROM golang:1.17-alpine
RUN apk add curl
...
```

Let's build the latest image by using `compose build`:

```
$ docker compose build
```

Adding health check to Compose

We have `curl` installed on the image; therefore, a health check will be added, utilizing the command:

```
services:
  task-manager:
    build:  .
    image: task-manager:0.1
    ports:
      - 8080:8080
    environment:
      - REDIS_HOST=redis:6379
    healthcheck:
      test: ["CMD", "curl", "-f", "http://localhost:8080/ping"]
      interval: 20s
      timeout: 10s
      retries: 5
      start_period: 5s
```

Every 20 seconds after an initial period of 5 seconds, a ping to `task-manager` will be issued using `curl` with a timeout of 10 seconds. In case of failure, there should be five retries before marking the instance *unhealthy*. By using `curl` with `-f`, there will be a failure on non-successful requests such as *5XX* responses.

If we run the application, we can observe from the logs that health check requests are being issued:

```
$ docker compose up
...
chapter2-task-manager-1  | [GIN] 2022/02/20 - 17:23:07 | 200 |
8.584µs |          127.0.0.1 | GET        "/ping"
chapter2-task-manager-1  | [GIN] 2022/02/20 - 17:23:27 | 200 |
34.459µs |          127.0.0.1 | GET        "/ping"
chapter2-task-manager-1  | [GIN] 2022/02/20 - 17:23:47 | 200 |
42.458µs |          127.0.0.1 | GET        "/ping"
```

On another shell, let's check the running containers:

```
$ docker ps
CONTAINER ID     IMAGE              COMMAND
CREATED           STATUS                          PORTS
NAMES
20c6d1cb557b     task-manager:0.1    "/task_manager"           4
seconds ago      Up 3 seconds (health: starting)    0.0.0.0:8080-
>8080/tcp    chapter2-task-manager-1
$ docker ps
20c6d1cb557b     task-manager:0.1    "/task_manager"           2
minutes ago      Up 44 seconds (healthy)    0.0.0.0:8080->8080/tcp
chapter2-task-manager-1
```

We can take it a bit further and set the endpoint to fail, so that we can see the service marked as unhealthy:

```
// Health Check
r.GET("/ping", func(c *gin.Context) {
        c.String(http.StatusInternalServerError, "pong")
})
```

After some time, the container is indeed marked unhealthy:

```
$ docker ps
CONTAINER ID     IMAGE              COMMAND
CREATED           STATUS                          PORTS
NAMES
20c6d1cb557b     task-manager:0.1    "/task_manager"           2
minutes ago      Up 2 minutes (unhealthy)    0.0.0.0:8080->8080/tcp
chapter2-task-manager-1
```

We investigated the health-check capabilities of Compose and managed to reproduce successful or failed health check scenarios.

Depending on services

The task-manager service depends on Redis. Provided the redis service is not up and running, task-manager will not be operational. For this case, we can instruct the service not to run until the service depending on it is fully running:

```
services:
  task-manager:
```

```
    build: .
    image: task-manager:0.1
    ports:
      - 8080:8080
    environment:
      - REDIS_HOST=redis:6379
    depends_on:
      - redis
```

ENTRYPOINT, arguments, and environment variables

Supposing that before we run the `task-manager` service, we also want to load the Redis database with some data. There are various ways to achieve this. Our choice would be to run another `redis` container using the **Redis CLI**. To do so, we need to override the *entry point* and specify some custom arguments.

Let's start with our initial service:

```
    redis-populate:
      image: redis
      depends_on:
        - redis
```

It is not functional for now, but we will build on it.

Environment files

Since we already used environment variables using the key-value method, we will try using a file. The environment variable file will contain the host and the port the Redis CLI should connect to:

```
  HOST=redis
  PORT=6379
```

The next step is to mount `env file` to the `redis-populate` service:

```
    redis-populate:
      image: redis
      depends_on:
        - redis
      env_file:
        - ./env.redis-populate
```

The script

Having a script that executes commands on the Redis CLI can be possible with a few tweaks. Suppose we want to add a task before starting the application; we would put the commands needed in a `txt` file:

```
HMSET task:a3a597d1-26f8-43e5-be05-81373f2c0dc3 Id a3a597d1-
26f8-43e5-be05-81373f2c0dc3 Name "Existing Task" Description "A
task that was here before" Timestamp 1645393434000
ZADD tasks 1645393434000 "a3a597d1-26f8-43e5-be05-81373f2c0dc3"
```

By using piping, we will actually forward the commands to be executed by the Redis CLI. The environment variable can also help to execute the commands to different servers. The script that will execute Redis commands from the `txt` file should look like this:

```sh
#!/bin/sh

cat $1| redis-cli -h $HOST -p $PORT
```

Those commands will be wrapped to an executable script and mounted to the `redis-populate` container:

```
redis-populate:
  image: redis
  depends_on:
    - redis
  env_file:
    - ./env.redis-populate
  volumes:
    - ./redis-populate.txt:/redis-populate.txt
    - ./redis-populate.sh:/redis-populate.sh
```

Custom ENTRYPOINT

In order to run the script and populate the database, we need to change `entrypoint` on Docker:

```
redis-populate:
  image: redis
  entrypoint: ["/redis-populate.sh","/redis-populate.txt"]
  depends_on:
    - redis
  env_file:
```

```
       - ./env.redis-populate
    volumes:
       - ./redis-populate.txt:/redis-populate.txt
       - ./redis-populate.sh:/redis-populate.sh
```

Thanks to the custom ENTRYPOINT and the environment variables, we should be able to have some data preloaded to our application.

Labels

Labels are a Docker feature that can be used on Compose. Through labels, we can apply metadata to Docker objects such as images, containers, networks, and volumes. This makes organizing the apps easier, for example, logical grouping or searching resources by using label filters.

Images

By building the image, we can specify a label:

```
    build:
      context: .
      labels:
         - com.packtpub.compose.app=task-manager
```

Then, we can filter related images using that label:

```
$ docker images --filter "label=com.packtpub.compose.app=task-manager"

REPOSITORY      TAG        IMAGE ID        CREATED         SIZE
task-manager    0.1        de4b8f7c9fff    15 hours ago    529MB
```

Containers

In the service section, we can add a tag to the container:

```
    redis:
      image: redis
      ports:
         - 6379:6379
      labels:
         - com.packtpub.compose.app=task-manager
```

Then, we can filter containers using that label:

```
docker images --filter "label=com.packtpub.compose.app=task-
manager"
```

```
$ docker ps --filter="label=com.packtpub.compose.app=task-
manager"
CONTAINER ID    IMAGE               COMMAND
CREATED         STATUS                      PORTS
NAMES
98c45777bdcd    redis               "docker-entrypoint.s…"    58
seconds ago    Up 57 seconds              0.0.0.0:6379->6379/tcp
chapter2-redis-1
```

We have been successful in packaging our application using Compose, creating the Docker image, and labeling it. This helped with running an application on Compose end-to-end, from the implementation of the code base until running it on a Docker container.

Summary

In this chapter, we created a multi-container application running on Docker Compose and used Compose features such as building and tagging an image, communicating between containers using the same network, health checks, labeling, and ENTRYPOINT configuration. Those features make it easier to transition from manual Docker-based management to a Compose-based one. By being able to package a complex application, we lay the foundation for deploying and distributing our multi-container application using Compose.

The next chapter is about networking and volume management, including understanding how the concept of Docker volumes and networks is mapped to Compose.

3

Network and Volumes Fundamentals

In the previous chapter, we managed to create our core **Go** application and provide storage using a Redis server. By having our core application in place, we proceeded to more advanced Docker concepts such as health checks, building images, tagging, and logically grouping using labels. The usage of **volumes** and **networking** was present throughout the chapter. Networking and volumes were being used all along; however, it was done transparently. This chapter will focus extensively on networks and volumes and how to configure them on a Compose-based application.

The first part will focus on volumes, how volumes map to Compose, how they work behind the scenes, and how to use them. Volumes play a crucial role in Docker. They allow you to attach external documentation and files that are needed for an application. Volumes can be shared and used to help with operations in an application.

The second part will focus on networks. Docker provides networking simply and transparently. In special scenarios, various options exist that can ensure multi-host communication, a different layer of networks between applications, and better isolation between containers.

In this chapter, we will cover the following main topics:

- Explaining Docker volumes
- Attaching a Docker volume to a container
- Docker volume drivers
- Declaring Docker volumes on Compose files
- Attaching a Docker volume to an existing application
- Creating a configuration file
- Mounting a file using volume
- Docker networking

- Defining networks on a Compose configuration
- Adding an extra network to the current application

Technical requirements

The code for the book is hosted on GitHub at `https://github.com/PacktPublishing/A-Developer-s-Essential-Guide-to-Docker-Compose`. In case of an update to the code, it will be updated on GitHub.

Explaining Docker volumes

When developing an application, a crucial aspect is to utilize operations that interact with a filesystem. Those operations can be driven by data storage purposes or, in some cases, might be configuration-driven. For example, take a database application that needs a filesystem to operate and store data. The same also applies to configurations that apply to an application. Let's imagine a JEE application that is heavy on *XML* configuration. It needs to be able to read and store data inside a filesystem. The issue with disk operations on a Docker container is that once the container is shut down, all of the changes will be lost. In this case, Docker provides us with volumes. Volumes are the preferred mechanism for persisting data that is generated and used by a container.

Here is a list of some of the characteristics of volumes:

- Volumes and their data can remain after a container has been deleted.
- A volume can be attached to another container.
- Volumes can be used by more than one container simultaneously.
- Volumes are abstract.
- Volumes can be local or remote.

Now that we have an overview of what volumes can provide, in the next section, we will proceed with adding volumes to containers and interacting with them.

Attaching a Docker volume to a container

We will create our first Docker volume using the following:

```
$ docker volume create example-volume
example-volume
```

We will inspect the volume that we've just created:

```
$ docker volume ls --filter="name=example-volume"
--format='{{json .}}'
{"Driver":"local","Labels":"","Links":"N/A","Mountpoint":"/var/
lib/docker/volumes/example-volume/_data","Name":"example-volume
","Scope":"local","Size":"N/A"}
```

By inspecting the preceding volume, we see that the volume is using the local driver. A local driver is the default driver option when creating volumes. The volume data will reside on the host that Docker Engine runs. There's also a mount point to our local filesystem. This is where the data in this volume resides physically on the host.

Since we created the volume, we will put it to use with a container. We will attach the volume and write some data on it. We can do this by mounting the volume to the container using --mount:

```
$ docker run -it --rm --name example-volume-container --mount
source=example-volume,target=/storage bash
bash-5.1# echo some-text > /storage/data-file.txt
bash-5.1# cat /storage/data-file.txt
some-text
bash-5.1# exit
```

Let's break down what has been executed into steps:

1. We run a bash container in interactive mode.
2. Once we exit the container, it will be stopped and deleted.
3. We mounted the volume at the /storage path.
4. We created a file on that path.
5. The file should be persisted on the volume.

We will create another container with the volume mounted and examine whether the file will be present:

```
$ docker run -it --rm --name second-volume-container --mount
source=example-volume,target=/storage bash
bash-5.1# cat /storage/data-file.txt
some-text
```

As expected, the volume is mounted, and the file is there.

In another Terminal session, we will inspect the container and the `Mounts` section:

```
$  docker inspect second-volume-container --format '{{json
.Mounts}}'
[{"Type":"volume","Name":"example-volume","Source":"/var/
lib/docker/volumes/example-volume/_data","Destination":"/
storage","Driver":"local","Mode":"z","RW":true,
"Propagation":""}]
```

Let's exit both Terminal sessions. As you can see, the volume has been attached and contains the volume information and the destination directory. After successfully creating a volume and attaching it to the containers, we will proceed with the scenario of a volume being shared.

Shared volumes

Earlier, we used one volume and attached it to a container. After stopping the container and mounting the volume to another container, we navigated to the directory and found that the data stored from the previous steps was there. A useful feature is that the volume can be attached to more than one container. By running the following commands in parallel, there will be writes on the same volume simultaneously:

```
$ docker run -it --rm   --name container-2 --mount
source=example-volume,target=/storage bash -c "for i in \$(seq
1 1 100000); do echo \$HOSTNAME \$i >> /storage/\$HOSTNAME.txt
; done"
$ docker run -it --rm   --name container-1 --mount
source=example-volume,target=/storage bash -c "for i in \$(seq
1 1 100000); do echo \$HOSTNAME \$i >> /storage/\$HOSTNAME.txt
; done"
```

As expected, if we attach the volume and navigate to the directory, we should see two different files created:

```
$ docker run -it --rm   --name container-1 --mount
source=example-volume,target=/storage bash
bash-5.1# ls /storage
770bdfc4dc94.txt   b9b09390cf17.txt data-file.txt
```

Both containers have been successful in storing files to the volume.

Read-only volumes

We were successful in using the original volume in a secondary container. There are scenarios where writes should be only executed from one container, whereas other containers could only read the data persisted by having *read-only* permissions.

As shown in the following example, mounting the volume as *read-only* prevents us from writing any files to it:

```
$ docker run -it --rm  --name read-only-volume-container
--mount source=example-volume,target=/storage,readonly bash
bash-5.1# cat /storage/data-file.txt
some-text
bash-5.1# touch /storage/test.txt
touch: /storage/test.txt: Read-only file system
bash-5.1# exit
```

So, we have become familiar with volumes and how to mount and use them with our containers. We also successfully mounted a volume for *read-only* purposes. In the next section, we will learn about volume drivers and what they can offer.

Docker volume drivers

We created Docker volumes and we put them to use in our containers. By inspecting the volume in the previous example, we can get some valuable information:

```
$ docker volume inspect example-volume
[
    {
        "CreatedAt": "2022-03-08T07:01:29Z",
        "Driver": "local",
        "Labels": {},
        "Mountpoint": "/var/lib/docker/volumes/example-volume/_
data",
        "Name": "example-volume",
        "Options": {},
        "Scope": "local"
    }
]
```

Here, "Driver" is set to "local"; this means that the volume data will reside locally in our workstation, and the driver will facilitate disk operations from the Docker container to the VM path.

We will use this as an opportunity to see the physical location of this file. In this example, which is running on macOS, we need to connect to the VM that hosts Docker Engine. We will do this using netcat:

```
$ nc -U ~/Library/Containers/com.docker.docker/Data/debug-
shell.sock
/ # cat /var/lib/docker/volumes/example-volume/_data/data-file.
txt
cat /var/lib/docker/volumes/example-volume/_data/data-file.txt
some-text
/ # exit
exit
```

Let's break down what we just did. On macOS and Windows, Docker Engine runs on a VM:

1. We need to shell to that VM and examine the directory.

2. Once we are inside the VM, we can confirm that the data is stored there.

As we understand that a volume can be located anywhere, the driver will facilitate making the usage of the volume possible.

Volume drivers can enable interactions with a remote filesystem and persist data. Available drivers include Azure File Storage, NFS, and AWS EFS.

Apart from highlighting the feature of interacting with a volume on a remote host, a volume driver can be used to add extra functionality when storing data; for example, by adding a form of proprietary encryption in transit or storage.

Using a volume driver versus mounting locally

Based on the contents of the last section, there are various options that we can use to interact with a remote filesystem. One option that is available and bypasses driver functionality is to mount the filesystem to the host. This way, the remote filesystem is attached to the container through the host that has already mounted it. It's up to the user and the use case to decide what should be used. Use cases can expand from having less maintenance overhead to the customized handling that a volume driver can provide.

So far, we have created some volumes using the command line and we also looked at how volumes work behind the scenes along with ways to interact with remote files using drivers. The next section will focus on using the volume knowledge that we have acquired in a Compose application.

Declaring Docker volumes on Compose files

We've been able to create a Docker volume and use it, and had a look at its features. The step next is to create and use volumes using Compose. Volumes on Compose are defined in the `volumes` sections:

```
services:
  nginx:
    image: nginx
volumes:
  example-compose-volume:
```

The configuration option on `volumes` can apply to *example labels*:

```
services:
  nginx:
    image: nginx
volumes:
  example-compose-volume:
    labels:
      - com.packtpub.compose.app=volume-example
```

Alternatively, you can have more advanced configurations, for example, a driver that is configured to use `nfs`:

```
services:
volumes:
  nfsvol:
    driver_opts:
      type: "nfs"
      o: "addr=127.0.0.1,nolock,rw"
      device: ":/data"
```

So, we have created and configured volumes using Compose. Now we can proceed with adapting our existing Compose application to use volumes.

Attaching Docker volumes to an existing application

The previous chapter was focused on a Task Manager application using a Redis database. Being a database, Redis is subject to backups and other disk operations.

Our previous Redis Compose configuration is as follows:

```
services:
  redis:
    image: redis
    ports:
      - 6379:6379
    labels:
      - com.packtpub.compose.app=task-manager
```

Redis has various options for persistence options (https://redis.io/topics/persistence). There are options from *point-in-time* to *AOF* (*Append-Only File*)-based persistence. We will focus on the point-in-time snapshot option that Redis provides, and we will use a volume to store those snapshots.

By default, .rdb backups on the Redis Docker image are stored within the /data directory. We will mount a volume to that location.

First, we will create the volume:

```
volumes:
  redis-data:
    labels:
      - com.packtpub.compose.app=task-manager
```

Then, we will attach the volume to the container:

```
services:
  redis:
    image: redis
    ports:
      - 6379:6379
    volumes:
      - redis-data:/data
```

Since the volume is attached to the database, the backups are stored in the volume and can also be accessed by another container. In the following section, we will create a custom configuration for the Redis service in order to be able to increase the frequency of our backups.

Creating a configuration file

Based on the documentation, we will enable Redis to take a snapshot of at least one key change in 60 seconds. Our Redis configuration will look like this:

```
dbfilename taskmanager.rdb
save 60 1
```

Here, we changed the backup name to `taskmanager.rdb`. We will take a snapshot every 60 seconds if at least one key changes. Now that the configuration is ready, we need a way to attach this configuration file to the database we created earlier using Compose.

Mounting a file using volume

The configuration file has been created and will be mounted by using **bind mounts**. A bind mount enables us to mount a file on the workstation running Docker Engine to the Docker container. When using Compose, bind mounts are defined in the `volumes` sections. This might be confusing since it's in the same section as `volumes`; however, it's a *bind mount*.

We will mount the Redis configuration file to the container and alter `entrypoint`:

```
services:
  redis:
    image: redis
    ports:
      - 6379:6379
    entrypoint: ["redis-server","/usr/local/etc/redis/redis.
conf"]
    labels:
      - com.packtpub.compose.app=task-manager
    volumes:
      - ./redis.conf:/usr/local/etc/redis/redis.conf
      - redis-data:/data
```

By inspecting the container, we can justify that the file has been mounted using bind mount:

```
$ docker inspect chapter3-redis-1 --format '{{json .Mounts }}'
...
        {
            "Type": "bind",
```

```
                    "Source": "/path/to/repo/A-Developer-s-
        Essential-Guide-to-Docker-Compose/Chapter3/redis.conf",
                    "Destination": "/usr/local/etc/redis/redis.
        conf",
                    "Mode": "rw",
                    "RW": true,
                    "Propagation": "rprivate"
                }
        . . .
```

Since we managed to configure Redis to take point-in-time snapshots on a volume of our choice, we will proceed with creating a backup of these snapshots using another process that has *read access* to the original backup volume.

Mounting read-only volumes

Now, we would like to save the snapshots created over time to a file. This way, we can restore the database from a point in time. However, we want to store the backups to another volume whose sole purpose will be for backups:

```
    . . .
        volumes:
            - ./redis.conf:/usr/local/etc/redis/redis.conf
            - redis-data:/data
    volumes:
        redis-data:
        backup:
```

Another container will be introduced. The container will have the responsibility of copying the Redis backups and moving them to the backup folder. The container will not be very sophisticated; it will just be a simple *bash script* that will run every few seconds.

The bash script for copying the Redis backups is as follows:

```
#!/bin/sh

while true; do cp /data/taskmanager.rdb /backup/$(date +%s
).rdb; sleep $BACKUP_PERIOD; done
```

The container will run the script that has been created. Since we've got everything we need, let's configure the container.

Both volumes need to be mounted. The Redis backup volume will be *read-only*. The responsibility of the backup process is to copy the current file on the Redis data directory and append a timestamp during the time of the backup:

```
redis-backup:
  image: bash
  entrypoint: ["/snapshot-backup.sh"]
  depends_on:
    - redis
  environment:
    - BACKUP_PERIOD=10
  volumes:
    - ./snapshot-backup.sh:/snapshot-backup.sh
    - redis-data:/data:ro
    - backup:/backup
volumes:
  redis-data:
  backup:
```

Note that we used the *read-only* option on data. This can also be defined alternatively:

```
volumes:
  - type: volume
    source: redis-data
    target: /data
    read_only: true
```

Our Redis database includes the snapshot backup process, as follows:

```
services:
  ...
  redis-backup:
    image: bash
    entrypoint: ["/snapshot-backup.sh"]
    depends_on:
      - redis
    environment:
      - BACKUP_PERIOD=10
    volumes:
```

```
        - ./snapshot-backup.sh:/snapshot-backup.sh
        - redis-data:/data:ro
        - backup:/backup
      labels:
        - com.packtpub.compose.app=task-manager
  volumes:
    redis-data:
      labels:
        - com.packtpub.compose.app=task-manager
    backup:
      labels:
        - com.packtpub.compose.app=task-manager
```

Let's run the preceding Compose application:

```
$ docker compose up
```

Now we should be able to execute the same cURL requests we executed in *Chapter 2, Running the First Application Using Compose.*

So far, we've been able to migrate our original Redis service on Compose to use a volume. Additionally, we've added another service to issue a backup process by having the backup volume mounted in *read-only* mode and storing the backups in a *backup purpose container*. In the next section, we will dive into Docker networking and learn how we can apply it to Compose.

Docker networking

A crucial feature of Docker is its networking features. The containers running on Docker communicate and interact transparently. Thanks to the network layer, Docker provides various networking aspects that are tackled efficiently, as follows:

- Private networks
- Internal DNS
- Container communication
- Bridging containers

Let's see the available networks that are currently running:

```
$ docker network ls
NETWORK ID      NAME            DRIVER      SCOPE
```

```
6a149c758fc2      bridge              bridge      local
9ec7758d7050      chapter1_default    bridge      local
7c95e79497b6      chapter2_default    bridge      local
16fda3c38e57      chapter3_default    bridge      local
4a6d6a3e8475      host                host        local
be28a5cd8a16      none                null        local
```

Networks on Docker can come in various forms. The most used formats include the following:

- Bridge
- Host
- Overlay

Let's look at each of them in the following sections.

Bridge

Bridge is the default network driver when using Docker. A bridge network on Docker is a software layer providing connectivity between containers that are connected to the same bridge network. Also, those containers are isolated from containers that are not connected to that network.

Previously when we inspected the network, we saw a network named `bridge`. This should be the default network that a container is attached to.

We will run an `nginx` container and check the network that the container will be attached to:

```
$ docker run --rm -p 8080:80 --name nginx-compose nginx
```

Now that the container is up and running, in another Terminal session, we will proceed with inspecting the network:

```
$ docker inspect --format '{{json .NetworkSettings.Networks }}'
nginx-compose
{"bridge":{"IPAMConfig":null,"Links":null,"Aliases":null,
"NetworkID":"6a149c758fc27c0c719b7c3c26b74c8f03c866ae1406ac
2902a1c45af720a79d","EndpointID":"4323e8f8bbceafcb74491e159
5073adc063fb424f87259ad1dd02078f59eeb13","Gateway":"172.
17.0.1","IPAddress":"172.17.0.2","IPPrefixLen":16,"IPv6Gateway"
:"","GlobalIPv6Address":"","GlobalIPv6PrefixLen":0,"MacAddress"
:"02:42:ac:11:00:02","DriverOpts":null}}
```

We can see that the container has been attached to the default bridge network, as we saw earlier:

```
$ NETWORK_ID=$(docker inspect --format '{{json .NetworkSettings
.Networks.bridge.NetworkID }}' nginx-compose|sed 's/"//g')
$  docker network ls --filter ID=$NETWORK_ID
NETWORK ID       NAME        DRIVER     SCOPE
6a149c758fc2     bridge      bridge     local
```

We can now exit the Terminal session by running the NGINX container using CTRL-C.

User-defined bridge network

A *user-defined* bridge network is superior to using the default network. Containers resolve each other's network location with a name or an alias:

- Containers are attached to a network bound for their application.
- Bridge network can be configured for the application needs.

Host

By using the Docker host, the container is not isolated from it. The container does not have its own IP address. The container will be accessed using the host IP.

We will run the nginx instance using the host network:

```
$ docker run --rm --network host --name host-nginx nginx
```

We will test that the server is running by using another container to access the container through localhost.

In another Terminal session, let's execute the following:

```
$ docker run -it --network host --rm --name nginx-wget bash
wget -O- localhost:80
Connecting to localhost:80 ([::1]:80)
writing to stdout
<!DOCTYPE html>
<html>
<head>
...
```

Here, instead of attaching the containers to a network, we mounted them to the host network. As expected, the container doesn't have an IP assigned; therefore, any requests to `localhost` are mapped to the host. This has made it feasible for the `nginx-wget` container to access the `host-nginx` container.

Overlay

The networks we investigated earlier centered on Docker Engine installed on one host. Using Docker on a production with an orchestration engine such as **Swarm** requires the use of multiple hosts. The overlay network creates a distributed network among those hosts. An *overlay* network will transparently join the host networks and create a unified network on top of them. Now that we have covered Docker networking and how it works, we will proceed to check how it works with Compose.

Defining networks on a Compose configuration

So far, our containers have been running and communicating transparently. This is because when we run a Compose application, a bridge network is created. Based on our previous examples, the containers have been able to interact with other Compose services and communicate through an alias:

```
$ docker network ls --filter name=chapter1_default
NETWORK ID      NAME                DRIVER    SCOPE
9ec7758d7050    chapter1_default    bridge    local
```

This can serve us for some time; however, we would like to have a different name for the network.

Adding a network in Compose is easy:

```
networks:
  task-manager-pubic-network:
    labels:
      - com.packtpub.compose.app=task-manager
```

However, creating a network is not enough. Containers need to be attached to the network. We will reiterate the application and the services to that network. With some adaption, our Compose file should have the services under the network:

```
// Chapter3/docker-compose.yaml
services:
  task-manager:
    build:
      context: ../Chapter2/.
      labels:
```

```
          - com.packtpub.compose.app=task-manager
      image: task-manager:0.1
      . . .
      networks:
        - task-manager-pubic-network
      labels:
        - com.packtpub.compose.app=task-manager
    redis:
      image: redis
      . . .
      networks:
        - task-manager-pubic-network
      labels:
        - com.packtpub.compose.app=task-manager
    redis-backup:
      image: bash
      . . .
      networks:
        - task-manager-pubic-network
      labels:
        - com.packtpub.compose.app=task-manager
  . . .
  networks:
    task-manager-pubic-network:
      labels:
        - com.packtpub.compose.app=task-manager
```

So, we managed to create a network for our Compose application and attach the services to it.

Next, we will put our application into action:

```
$ docker compose up
```

Since our service has the network attached, we will inspect the existing Docker networks:

```
$ docker network ls
NETWORK ID      NAME                               DRIVER
SCOPE
```

```
6a149c758fc2    bridge                                        bridge
local
ce5aa144f0f9    chapter3_task-manager-pubic-network           bridge
local
4a6d6a3e8475    host                                          host
local
be28a5cd8a16    none                                          null
local
```

Since all our containers have been attached to the newly created network, they are under the hood of one network. The default Compose network that was created earlier is not there.

Adding an extra network to the current application

So far, our containers have been running and communicating transparently, and we have managed to define the bridge network the services would use. Another thing we have observed is that the bridge network that used to be created by default no longer exists. Since we saw only one network being used at a time, we will add multiple networks in Compose.

The target for this would be our Redis database. In real life, databases usually reside on another network and are linked to your application:

```
networks:
  redis-network:
    labels:
      - com.packtpub.compose.app=task-manager
```

We will attach the Redis database to that network:

```
redis:
    image: redis
    ports:
      - 6379:6379
    entrypoint: ["redis-server","/usr/local/etc/redis/redis.
conf"]
    volumes:
      - ./redis.conf:/usr/local/etc/redis/redis.conf
      - redis-data:/data
      - backup:/backup
    networks:
```

```
    - redis-network
  labels:
    - com.packtpub.compose.app=task-manager
```

If we are curious and try to run the application, we will get an error:

```
{ "id": "8b171ce0-6f7b-4c22-aa6f-8b110c19f838", "name": "A
task", "description": "A task that need to be executed at the
timestamp specified", "timestamp": 1645275972000 }
```

As expected, the Go application does not have access to the network that the Redis service is attached to. We will fix this by attaching the task-manager application to the Redis network:

```
  task-manager:
    build:
      context: ../Chapter2/.
      labels:
        - com.packtpub.compose.app=task-manager
    image: task-manager:0.1
    ports:
      - 8080:8080
    environment:
      - REDIS_HOST=redis:6379
    depends_on:
      - redis
    healthcheck:
      test: ["CMD", "curl", "-f", "http://localhost:8080/ping"]
      interval: 10s
      timeout: 5s
      retries: 5
      start_period: 5s
    networks:
      - task-manager-pubic-network
      - redis-network
```

So, the container can be attached to two networks. By attaching the container to the Redis network, communication is made possible.

If we run a GET request, we should see a positive result, as follows:

```
curl --location --request GET 'localhost:8080/task/8b171ce0-
6f7b-4c22-aa6f-8b110c19f83a'
{"task":{"id":"8b171ce0-6f7b-4c22-aa6f-8b110c19f83a","name":"A
task","description":"A task that need to be executed at the
timestamp specified","timestamp":1645275972000}}
```

Summary

In this chapter, we provisioned volumes, attached them to Compose services, and shared them between services for maintenance operations. We identified the usages of networks in Docker and how they fit into our Compose usage scenarios. By using these networks, we analyzed how a network works on Compose and how we can have multiple networks running in Compose applications.

The next chapter will focus on the Docker Compose commands. We will dive into the available commands, their purpose, and how they can assist us in managing and provisioning Compose applications.

4

Executing Docker Compose Commands

In the previous chapter, we focused on using Docker networks and volumes on a Compose application. By using networks, we managed to establish connectivity between the components of a Compose application; by using volumes, we facilitated I/O operations and kept the data created portable and permanent.

So far, we have used various Docker Compose commands in order to provision our applications and interact with Compose components. This chapter will focus on the available Compose commands and their options. Once we have an overview of the available commands, we will dive into the provisioning commands and the commands interacting with the Compose containers. After that, we will learn more about cleanup commands. By covering the commands that assist in our application development, we will have a deep dive into monitoring commands.

The following main topics will be covered in this chapter:

- Introducing Compose commands
- The Docker CLI versus Compose commands
- Provisioning commands
- Container commands
- Cleanup commands
- Image commands
- Monitoring commands
- Other commands

Technical requirements

The code for the book is hosted on GitHub at `https://github.com/PacktPublishing/A-Developer-s-Essential-Guide-to-Docker-Compose`. If there is an update to the code, it will be updated on the GitHub repository.

Introducing Compose commands

Compose has a variety of commands that can be used to interact with containers, monitor applications, and also to manage images.

Throughout this chapter, we shall examine Compose commands, apply them to a target application, break down their usage, and understand how they work behind the scenes.

The commands we will focus on are the following:

- `build`
- `create`
- `up`
- `images`
- `pull`
- `push`
- `down`
- `rm`
- `logs`
- `top`
- `ps`
- `events`

We will learn more about these commands, see how we can benefit from them, and utilize them on a Compose application.

The Docker CLI versus Compose commands

Most of these commands seem familiar to the ones we used through the Docker CLI. The difference between the Compose and the Docker CLI commands is that the CLI commands operate through all the Docker components in your system, whereas the Compose ones are limited in the context of the Compose application specified in the `docker-compose.yaml` file.

Behind the scenes, the `Compose` command will parse the Compose file and retrieve information about the application. The API calls issued to the Docker Engine will contain filters based on the information retrieved previously and will interact only with the services and resources of the application. For example, the `ps` command, instead of retrieving all the running containers, which would have happened if we used the Docker CLI, will retrieve only the containers that should run on the Compose application.

Setting up the target application

In order to make the commands showcase feasible, we will combine the work of the previous chapters and assemble a Compose application. The application should use the source code of the Task Manager application we developed in *Chapter 2, Running the First Application Using Compose*, along with a Redis database.

The Compose service definition uses the previous chapter's code:

```
services:
  task-manager:
    build:
      context: ../Chapter2/.
    image: docker.io/library/chapter4_task-manager
    ports:
      - 8080:8080
    environment:
      - REDIS_HOST=redis:6379
  redis:
    image: redis
    ports:
      - 6379:6379
```

We have a summary of the commands that we will focus on and also an application set up in order to apply those commands. In the next section, we shall focus on commands that provision resources.

Provisioning commands

By having a Compose application file set up, we can now provision the application. The commands we will use are the following:

- `build`
- `create`
- `up`

Let's take a look at them.

build

As seen previously, we do not need to build a Docker image and tag it manually before using it on Compose. Compose can build the image and use it for the application.

Issuing the `build` command creates the image:

```
$ docker compose build
=> => exporting layers
0.0s

...

=> => naming to docker.io/library/chapter4_task manager
0.0s
$ docker images
REPOSITORY                          TAG         IMAGE ID
CREATED             SIZE
chapter4_task-manager               latest      58247e548811    35
hours ago       529MB
```

By using the `build` command, we create the image for the Task Manager application, and it has been tagged appropriately. We shall now proceed with getting the application up and running.

create

`create` has been deprecated, since we can use up instead. The purpose of `create` was to create the containers, volumes, and networks but without starting the containers. This functionality has been replaced by up `-no-start`:

```
$ docker compose create
[+] Running 3/0
⊞ Network chapter4_default          Created
0.0s
⊞ Container chapter4-redis-1        Created
0.0s
⊞ Container chapter4-task-manager-1 Created
```

up

up has many usages and parameters. We will replace the previous `create` operations with up:

```
$ docker compose up --no-start
[+] Running 3/3
⊞ Network chapter4_default          Created
0.0s
⊞ Container chapter4-task-manager-1  Created
0.0s
⊞ Container chapter4-redis-1         Created
$ docker ps -a
CONTAINER ID    IMAGE                                        COMMAND
CREATED          STATUS                      PORTS       NAMES
3d1ad2f353d3    redis                                        "docker-
entrypoint.s…"   3 seconds ago    Created
chapter4-redis-1
b6d1a5f16313    chapter4_task-manager                        "/
task_manager"            3 seconds ago    Created
chapter4-task-manager-1
```

The containers and networks have been created but do not run.

Since we created the resources, we will start the application. A useful command is -d or –detach, the detached mode. The detached mode can help run the Compose application without being bound to a terminal:

```
$ docker compose up --detach
[+] Running 2/2
⊞ Container chapter4-task-manager-1  Started
0.3s
⊞ Container chapter4-redis-1         Started
```

As we have previously created the resources and built the images, we will instruct up not to create the image by using --no-build:

```
$ docker compose up --no-build --detach
[+] Running 3/3
⊞ Network chapter4_default          Created
0.0s
```

```
⊞ Container chapter4-redis-1          Started
0.3s
⊞ Container chapter4-task-manager-1   Started
```

If we need to update a service, we can update and delete the resources from old services. For example, in the case of rolling out task-manager-2, we would like the old container versions to be removed.

Let's suppose we renamed the service as follows:

```
services:
  task-manager-2:
. .
```

If we execute up, a warning shall be printed:

```
$ docker compose up
WARN[0000] Found orphan containers ([chapter4-task-
manager-2-1]) for this project. If you removed or renamed this
service in your compose file, you can run this command with the
--remove-orphans flag to clean it up.
```

We shall tackle this by using --remove-orphans:

```
$ docker compose up --remove-orphans
[+] Running 3/0
⊞ Container chapter4-task-manager-1      Removed
0.0s
⊞ Container chapter4-redis-1            Created
```

We've managed to get the application up and running by using the provisioning commands shown in this chapter. We can now use Compose commands to interact with the containers that the Compose services are running upon; those commands will solely be focused on interacting with containers.

Container commands

Container commands enable us to start, stop, and restart containers, execute commands upon them, and also kill them.

exec

Throughout the book, we have executed commands using the Docker `exec` command. Docker Compose also provides this functionality. The difference is that instead of running the `exec` command directly to a container on Compose, we specify the service to execute upon. The command then will be routed to the service's underlying container:

```
$ docker compose exec task-manager ls
go.mod     go.sum     main.go
```

If we check the existing containers, we will not see any extra containers provisioned:

```
$ docker ps -a
CONTAINER ID     IMAGE                                    COMMAND
CREATED            STATUS                        PORTS
NAMES
8e559f9bf6d1     chapter4_task-manager                   "/
task_manager"              18 seconds ago    Up 17 seconds
0.0.0.0:8080->8080/tcp    chapter4-task-manager-1
46ca9a239557     redis                                    "docker-
entrypoint.s..."    18 seconds ago    Up 17 seconds
0.0.0.0:6379->6379/tcp    chapter4-redis-1
```

run

`run` might seem similar to `exec`, but it differs in the fact that a new container will spin up. This makes it feasible to run a container in an environment provisioned by a Compose file. Thus, the container will have access to the volumes and networks that have been created:

```
$ docker compose run task-manager sh
/app #
```

On another terminal, we should inspect whether the container is running:

```
$ docker ps -a
CONTAINER ID     IMAGE
COMMAND                       CREATED            STATUS
PORTS                         NAMES
b7920b959c98     host.docker.internal:5000/task-manager:0.1
"sh"                          23 minutes ago    Up 23 minutes
8080/tcp                      chapter4_task-manager_run_7ad8bcc01122

...
```

```
b575f8bcd7d3    host.docker.internal:5000/task-manager:0.1
"/task_manager"              2 hours ago       Up 42 minutes
0.0.0.0:8080->8080/tcp    chapter4-task-manager-1
```

As shown, a new container called chapter4_task-manager_run_7ad8bcc01122 has been
created; by exiting, the container will be stopped.

pause

On Docker, the pause command suspends the processes of the specified container. By using pause
on Compose, we pause the containers of the service specified:

```
$ docker compose pause task-manager
[+] Running 1/0
⊞ Container chapter4-task-manager-1  Paused
docker ps
CONTAINER ID    IMAGE                    COMMAND
CREATED         STATUS                   PORTS
NAMES
8e559f9bf6d1    chapter4_task-manager    "/task_manager"
7 minutes ago   Up 7 minutes (Paused)    0.0.0.0:8080->8080/tcp
chapter4-task-manager-1
46ca9a239557    redis                    "docker-entrypoint.s..."
7 minutes ago   Up 7 minutes             0.0.0.0:6379->6379/tcp
chapter4-redis-1
```

unpause

Since we previously paused the task-manager container, we shall unpause the container using
the unpause command:

```
$ docker compose unpause task-manager
[+] Running 1/0
⊞ Container chapter4-task-manager-1  Unpaused
0.0s
$ docker ps
CONTAINER ID    IMAGE                    COMMAND
CREATED         STATUS       PORTS                    NAMES
8e559f9bf6d1    chapter4_task-manager    "/task_manager"
8 minutes ago   Up 8 minutes    0.0.0.0:8080->8080/tcp
chapter4-task-manager-1
```

```
46ca9a239557    redis                   "docker-entrypoint.s…"
8 minutes ago    Up 8 minutes    0.0.0.0:6379->6379/tcp
chapter4-redis-1
```

As expected, the container is unpaused.

start and stop

By using `stop`, we should stop running containers. `start` can be used to start existing containers. If a container is stopped, it will not be removed. We will shell into a container, create a file, and then stop the container. Once we start, the file should be there:

```
$ docker compose ps
NAME                        COMMAND              SERVICE
STATUS                 PORTS
chapter4-task-manager-1    "/task_manager"      task-manager
running                 0.0.0.0:8080->8080/tcp
$ docker compose exec task-manager sh -c "echo test > text.txt"
$ docker compose stop task-manager
[+] Running 2/0
⊞ Container chapter4-task-manager-1                   Stopped
0.1s
```

The container is stopped. We expect that by starting the container, the file we created previously will exist:

```
$ docker compose start task-manager
[+] Running 1/1
Container chapter4-task-manager-1  Started
$ docker compose exec task-manager sh -c "cat text.txt"
test
```

restart

`restart` will restart the Compose service. Even if the service is stopped, it will start:

```
$ docker compose restart task-manager
[+] Running 1/1
 Container chapter4-task-manager-1  Started
```

Let's stop a service and see whether it will start:

```
$ docker compose stop task-manager
[+] Running 1/0
Container chapter4-task-manager-1   Stopped
$ docker compose restart task-manager
[+] Running 1/1
Container chapter4-task-manager-1   Started
0.2s
```

Regardless of the previous state of the application, the application started.

kill

kill can be used to kill the containers of a service:

```
$ docker compose kill task-manager
```

ps

compose ps is like docker ps. The key difference is that the target of the compose ps is filtered based on the ones specified in the Compose file:

```
$ docker compose ps
NAME                          COMMAND           SERVICE
STATUS                 PORTS
chapter4-task-manager-1    "/task_manager"      task-manager
running                0.0.0.0:8080->8080/tcp
```

We can display only the running services with the --services option:

```
$ docker compose ps --services
task-manager
```

We can display only the container ID using --quiet:

```
$ docker compose ps --quiet
9b612d9ed210ab96d2b340ed8840781c80288097e1af6a96ed208d6b6d1
fb42a
```

We can even show stopped containers – for example, if we run a command once on the task-manager service:

```
$ docker compose run task-manager sh
/app # exit
```

Using ps -a, we can show the stopped container, created by the run command:

```
$ docker compose ps -a
NAME                                        COMMAND
SERVICE                 STATUS              PORTS
chapter4-task-manager-1                      "/task_manager"
task-manager            running             0.0.0.0:8080->8080/tcp
chapter4_task-manager_run_91e7eec48f5d    "sh"
task-manager            exited (0)
```

In this section, we created our base Compose application. By using Compose commands, we have been able to build the application images the application required. By using commands to run the application, we also evaluated the extra options the commands offer. Then, we used Compose commands to interact with the containers spun up by running the application. We managed to execute commands to existing containers as well as to spin up containers using the environment's existing resources. In the next section, we will cover commands that clean up the resources we provisioned using Compose.

Cleanup commands

A common case with Docker is the various resources that are created when using it for day-to-day development. Removing images, containers, and also networks and volumes can become redundant. Compose offers a more managed way to deal with all those resources.

down

Let's run the application in attached mode:

```
$ docker compose up
[+] Running 2/0
⊞ Container chapter4-task-manager-1 Running                    0.0s
⊞ Container chapter4-redis-1 Running                           0.0s
```

Once we hit *Ctrl + D*, we shall escape Compose, the logs will stop being displayed, and the services shall stop running:

```
Gracefully stopping... (press Ctrl+C again to force)
[+] Running 2/2
⊞ Container chapter4-task-manager-1   Stopped
0.1s
⊞ Container chapter4-redis-1          Stopped
0.1s
canceled
```

As indicated by the messages, the application has stopped. The resources that have been provisioned are still there. We can examine that using `images` to see the images created, as well as `ps` to see containers that have been running:

```
$ docker compose images
Container                 Repository             Tag
Image Id           Size
chapter4-redis-1          redis                  latest
f16c30136ff3       107MB
chapter4-task-manager-1   chapter4_task-manager  latest
58247e548811       529MB
$ docker compose ps
NAME                      COMMAND                SERVICE
STATUS             PORTS
chapter4-redis-1          "docker-entrypoint.s…" redis
exited (0)
chapter4-task-manager-1   "/task_manager"        task-manager
exited (2)
$ docker network ls
NETWORK ID      NAME                                DRIVER
SCOPE
...
ed28fb3286ab    chapter4_default                    bridge
local
```

By running `docker compose` up again, the same resources will be used.

However, we might want to remove those resources from our system and free some space, or perhaps the application that we have been developing is redundant. In this case, we can use docker compose down. This will remove the resources provisioned by the Compose application:

```
$ docker compose down
[+] Running 3/2
⊞ Container chapter4-task-manager-1    Removed
0.1s
⊞ Container chapter4-redis-1           Removed
0.1s
⊞ Network chapter4_default             Removed
0.0s
$ docker compose ps
NAME                    COMMAND              SERVICE
STATUS                  PORTS
$   docker network ls|grep chapter4
```

Based on the commands run afterward, it seems that all the resources have been removed. However, there is still something that has been created while running the application, which is the images we built previously using docker compose build.

down comes with an option to remove the image that was built when running the application:

```
$ docker compose down --rmi local
[+] Running 4/2
⊞ Container chapter4-redis-1           Removed
0.2s
⊞ Container chapter4-task-manager-1    Removed
0.2s
⊞ Image chapter4_task-manager         Removed
0.0s
⊞ Network chapter4_default            Removed
0.0s
```

local will work for the images that have been built using the compose file and don't have an image name. In our case, since we do have a tag, image: chapter4_task-manager:latest, we need to use the all parameter:

```
$ docker compose down --rmi all
[+] Running 5/0
```

▦ Container chapter4-task-manager-1	Removed
0.0s	
▦ Container chapter4-redis-1	Removed
0.0s	
▦ Image redis	Removed
0.0s	
▦ Image docker.io/library/chapter4_task-manager	Removed
0.0s	
▦ Network chapter4_default	Removed
0.0s	

Be aware that `all` will also remove the locally cached image, as we can see in the preceding code.

rm

In the previous steps, the usage of `down` removed all the containers. By using `rm`, we can remove containers individually. Let's suppose we have the scenario of a poorly running container:

```
services:
  failed-manager:
    build:
      context: ../Chapter2/.
    entrypoint: ["/no-such-command"]
  task-manager:
    build:
      context: ../Chapter2/.
    image: chapter4_task-manager:latest
    ports:
      - 8080:8080
    environment:
      - REDIS_HOST=redis:6379
  redis:
    image: redis
    ports:
      - 6379:6379
```

The `failed-manager` service will fail:

```
$ docker compose up -d
[+] Running 3/4
```

```
⊞ Network chapter4_default            Created
  0.0s
⊞ Container chapter4-failed-manager-1  Starting
  0.5s
⊞ Container chapter4-redis-1          Started
  0.4s
⊞ Container chapter4-task-manager-1   Started
  0.4s
Error response from daemon: OCI runtime create failed:
container_linux.go:380: starting container process caused:
exec: "/no-such-command": stat /no-such-command: no such file
or directory: unknown
```

We do see the error. If we use ps, we should see that one container is running and that the other container was just created:

```
$ docker compose ps
NAME                       COMMAND                SERVICE
STATUS              PORTS
chapter4-failed-manager-1  "/no-such-command"     failed-
manager      created               8080/tcp
chapter4-redis-1           "docker-entrypoint.s…" redis
running             0.0.0.0:6379->6379/tcp
chapter4-task-manager-1    "/task_manager"        task-
manager         running            0.0.0.0:8080->8080/tcp
```

Let's remove the bad container:

```
$ docker compose rm failed-manager
? Going to remove chapter4-failed-manager-1 Yes
[+] Running 1/0
⊞ Container chapter4-failed-manager-1  Removed
```

Alternatively, we can force a removal using the -f or --force flag:

```
$ docker compose rm failed-manager --force
Going to remove chapter4-failed-manager-1
[+] Running 1/0
⊞ Container chapter4-failed-manager-1  Removed
```

By using force, we avoided Command Prompt.

So far, after managing to run an application and interact with the containers that were created, we proceeded to stop the application, as well as to clean up the resources created, in order to make the running application feasible. One resource that has been created through the build process was the application's image. Now, we shall dive deeper into managing images using Compose.

Image commands

Compose provides us with various options when using images. This makes its usage efficient, since there is no need to build images individually and use them in an application. Building images, tagging them, and also pushing them to a repository is something that can be done through Compose. This leads to development efforts being focused in one place.

List images

Our application so far has two services backed by containers. We can see the images used by the containers using the docker compose images command:

```
$ docker compose images
Container                    Repository                  Tag
Image Id              Size
chapter4-redis-1             redis                       6.2.6
23d787aaa419          107MB
chapter4-task-manager-1   chapter4_failed-manager     latest
58247e548811          529MB
```

We can also make the output less verbose by using the --quiet option:

```
$ docker compose images --quiet
58247e548811b8812d48467436bc07ed40b4a7d4cd8328e57234d465ef189
14a
23d787aaa419ab884dd8682dca3153506f4dd00aba2ca9cd5953e94cae36b
d7d
```

Pulling images

By using the images command, we have been able to retrieve the images used by the Compose application. What if we changed the image? The expected outcome would be to pull the image before running. Compose provides the pull command, which should pull the image specified.

So, let's specify another `redis` image:

```
services:
  redis:
    image: redis:6.2.6
```

Instead of using up and pulling the image by default, we will use `pull`:

```
$ docker compose pull --ignore-pull-failures --parallel
[+] Running 2/3
⊞ failed-manager Skipped                        0.0s
⊞ redis Pulled                                  1.2s
⊞ task-manager Error                            1.5s
Pulling task-manager: Error response from daemon: pull access
denied for chapter4_task-manager, repository does not exist or
may require 'docker login': denied: requested access to the
resource is denied
$ echo $?
0
```

As you can see, by issuing the `pull` command, we pulled the older Redis images specified. Also, because the task manager does not exist as an image in the registry and is a local one, it failed to pull. This did not block the operation; thus, the return of the command was 0. We achieved this by using the `--ignore-pull-failures` option. Also, by using the `--parallel` option, the images are pulled in parallel.

Pushing images

Pushing images to a Docker registry can also be done using Compose. To show this example, we will set up a registry using Compose.

Local Docker registry on Compose

In order to push Docker images, a registry is required. We will deploy a registry using Compose and configure our Compose application to use this registry.

Our Compose registry file is as follows:

```
services:
  registry:
    image: registry:2
```

```
    ports:
      - 5000:5000
```

The registry will run on port 5000; thus, we need to make sure this port is available on our system. Keep in mind that this is not a secure registry, so we need to edit the Docker daemon config to explicitly specify the non-secure registry.

The location of the daemon.json config varies on the three main operating systems:

- **Linux**: /etc/docker/daemon.json

- **Windows**: C:\ProgramData\docker\config\daemon.json

- **macOS**: ~/.docker/daemon.json

Once the file has been located, we need to add the following entry:

```
    "insecure-registries" : [
      "host.docker.internal:5000"
    ],
```

After editing, the file should look like this:

```
    {
    ...
      "insecure-registries" : [
        "host.docker.internal:5000"
      ],
      "builder" : {
    ..
      }
    }
```

Pushing to the local registry

Now, it is feasible to push the task-manager image to the local registry, although any available registry can work. Since the registry is a local one run by the Docker engine, we will point to it by using the host.docker.internal DNS:

```
    services:
      task-manager:
        build:
```

```
      context: ../Chapter2/.
    image: host.docker.internal:5000/task-manager:0.1
    ports:
      - 8080:8080
    environment:
      - REDIS_HOST=redis:6379
  redis:
    image: redis
    ports:
      - 6379:6379
```

We can now build the image:

```
$ docker compose build
[+] Building 2.0s (17/17) FINISHED

=> [internal] load build definition from Dockerfile
...
=> => naming to host.docker.internal:5000/task-manager:0.1
...
```

Then we push the image:

```
$ docker compose push
 [+] Running 1/13
⊞ redis Skipped
0.0s
⊞ Pushing task-manager: 98ca4aa0fc4c Pushed
15.8s
...
⊞ Pushing task-manager: 590efbee44c0 Pushed
15.8s
```

We built the images using Compose, and we also managed to push to a registry we created. We inspected the images used and pulled the new images specified in the compose file. The next section will be focused on monitoring our Compose application.

Monitoring commands

A common case in day-to-day development is monitoring an application and ensuring it is operating properly. From an application using too many resources to an elusive bug, or even an application restarting, there are various cases of an application malfunctioning. For these cases, monitoring commands play a crucial role.

Logs

Logs provide the ability to view the logs of a Compose application running on our system. If we run the up command in detached mode, we won't be able to see any logs:

```
$ docker compose up -d
[+] Running 2/2
⊞ Network chapter4_default          Created
0.0s
⊞ Container chapter4-task-manager-1  Started
```

However, the logs do get generated from our applications; thus, we can retrieve them:

```
$ docker compose logs
chapter4-task-manager-1  | [GIN-debug] [WARNING] Creating an
Engine instance with the Logger and Recovery middleware already
attached.
...
chapter4-task-manager-1  | [GIN-debug] Listening and serving
HTTP on :8080
```

What logs did was display all the logs the application produced on the terminal and exit. There is also the option of --follow. This way, we can follow the log output and have all new logs printed on our screen:

```
$ docker compose logs -f
chapter4-task-manager-1  | [GIN-debug] [WARNING] Creating an
Engine instance with the Logger and Recovery middleware already
attached.
...
chapter4-task-manager-1  | [GIN-debug] Listening and serving
HTTP on :8080
```

Apart from that, there are options such as --no-color, where no color is used on the terminal; --timestamps, which displays a timestamp of the log when captured; and --tail, which limits the logs displayed from each container:

```
$ docker compose logs -f --timestamps --tail="3" --no-color
chapter4-task-manager-1  | 2022-03-26T07:50:13.846624301Z
[GIN-debug] [WARNING] You trusted all proxies, this is NOT
safe. We recommend you to set a value.
chapter4-task-manager-1  | 2022-03-26T07:50:13.846625967Z
Please check https://pkg.go.dev/github.com/gin-gonic/
gin#readme-don-t-trust-all-proxies for details.
chapter4-task-manager-1  | 2022-03-26T07:50:13.846627551Z
[GIN-debug] Listening and serving HTTP on :8080
```

top

logs is indeed a useful command, but it is limited only to the logs and the log messages that an application emits. By using top, we display the running Compose processes for the file present:

```
$ docker compose top
chapter4-task-manager-1
UID     PID    PPID   C    STIME    TTY    TIME        CMD
root    7913   7887   0    07:50    ?      00:00:00    /task_
manager
```

top on Docker is limited to the container specified. When using top on Compose, it will monitor the containers listed in the docker-compose.yaml file.

Events

events is a functionality of Compose similar to the events command that Docker provides. By running the events command, we listen for events that happen on the Compose application:

```
$ docker compose events
...
```

Let's stop the container:

```
$ docker stop chapter4-task-manager-1
chapter4-task-manager-1
```

On the `events` tab, we should see the deletion:

```
$ docker compose events
...
2022-03-26 10:05:45.233256 container stop
2cda5390558dc9d6b12bae95d6c65b23a6d95b36a5279677834f81af1d62
69f2 (name=chapter4-task-manager-1, com.packtpub.compose.app
=task-manager, image=host.docker.internal:5000/task-
manager:0.1)
```

Every operation that produces Docker `events` commands and affects the components of our application will be printed on the screen.

We just covered monitoring commands; thus, we are now able to track events of our application, identify the resources being used, as well as to troubleshoot application issues using logs.

Other commands

There are commands that can help in printing the available information.

help

`help` is a terminal command that provides usage instructions for commands available.

version

As seen in previous chapters, `version` displays the currently running version of Compose:

```
$ docker compose version
Docker Compose version v2.2.3
```

port

`port` prints the port for a binding:

```
$ docker compose port task-manager 8080
0.0.0.0:8080
```

config

`config` is used to validate the Compose configuration and instruct information to print:

```
$ docker-compose config
services:
  redis:
    image: redis
    ports:
    - published: 6379
      target: 6379
  task-manager:
    build:
      context: /path/to/Chapter2
    environment:
      REDIS_HOST: redis:6379
    image: host.docker.internal:5000/task-manager:0.1
    ports:
    - published: 8080
      target: 8080
version: '3.9'
```

It can also be used to display only the services:

```
$ docker-compose config --services
task-manager
redis
```

Alternatively, it can display the volumes specified:

```
$ docker-compose config --volumes
```

By using the preceding commands and learning more about their usage, we now have a complete overview of Compose commands that are useful on a day-to-day basis.

Summary

In this chapter, we had an extensive look at the available Compose commands and their options. We used commands that helped us to provision our multi-container application, as well as commands that assist us to interact with the containers of our application. Thanks to the image functionalities Compose provides, we pulled, built, and deployed images to a Docker registry without issuing any non-Compose commands. By managing to run our applications and evaluate the existing command options, we also proceeded to monitor our application, either by monitoring the logs or listening to Docker events, and we even monitored the activity of our Compose processors using `top`.

The following chapters will move on to more specific concepts of Compose that can benefit our daily development. We will transition to a microservice-based application using Compose, and monitor it, modularize it, and build it using CI/CD.

Part 2:
Daily Development with Docker Compose

This part will take us through more advanced concepts of Docker Compose. We shall set up databases for daily usage using Compose. Then, using Docker, networking microservices will communicate with each other. We will run entire stacks locally on Compose and simulate production environments. Lastly, we shall enhance our CI/CD jobs using Docker Compose.

The following chapters will be covered under this section:

5
Connecting Microservices

The previous part was focused on getting started with Docker. Once we had installed Docker on our workstation, we learned more about Docker Compose and its day-to-day usage. We learned to combine Docker images using Compose and running multi-container applications. After successfully running multi-container applications, we moved on to more advanced concepts, such as Docker volumes and networks. Volumes helped to define how to store and share data, while networks made it possible to isolate certain applications and access them only through a specific network. During this process, we gradually moved away from using Docker CLI commands to Docker Compose commands. By using Compose commands, our focus shifted to the Compose application we provisioned, and it was possible to interact with the containers, monitor and execute administrative commands upon them, and focus our operations on the resources provisioned by Compose.

Since we have already created multi-container applications using Compose, we can now proceed with more advanced scenarios involving Compose. Nowadays, applications have become more complex. This leads to the need to split an application into multiple applications, either for scaling or team purposes. Microservices are the new norm. By splitting a problem into smaller parts, teams can benefit by increasing their delivery rate. Also, microservices can help with performance tuning and scale the parts of an application that matter the most.

Although the concept of microservices existed long before the introduction of Docker, it played a crucial role in mass microservice adoption. The way services can be isolated and deployed everywhere, packaged with the tools needed, made it possible to reduce the cost and effort of deploying a microservice to a virtual or physical machine.

By using microservices, an application is split into multiple parts. Communication between the services is crucial. There are public-facing microservices, the entry points of an application, and microservices that are only internal.

In this chapter, we will focus on the application introduced in *Chapter 2, Running the First Application Using Compose*, the Task Manager application, and transform it into a microservice-based application. We will introduce a microservice, the geolocation service, which will be used by the Task Manager. By introducing this service, we will add it to a network that can be accessed only by the Task Manager application. After that, another service will be introduced, which will generate analytics based on data streamed to Redis.

Overall, this chapter will focus on the following topics:

- Introducing the location microservice
- Adding the location microservice to Compose
- Adding the location microservice to a private network
- Executing requests to a location microservice
- Streaming task events
- Adding a task events processing microservice

Technical requirements

The code for the book is hosted on GitHub at `https://github.com/PacktPublishing/A-Developer-s-Essential-Guide-to-Docker-Compose`. If there is an update to the code, it will be updated on the GitHub repository.

Introducing the location microservice

By using the Task Manager application, introduced in *Chapter 2, Running the First Application Using Compose*, we will enhance its functionality by adding a location where a task should take place. Each task will have a location. By gathering those tasks, the locations will be stored; thus, each time a task is created, locations that have been previously visited will be recommended.

We shall create the location service as a new microservice. The service will not share anything with the Task Manager. It will have an API of its own. For simplicity, we shall use the same programming language we used previously, Golang, as well as the same database, Redis.

Let's proceed with a Redis instance. Since will we use Compose, the following will be our configuration:

```
services:
  redis:
    image: redis
```

We can run in detached mode:

```
$ docker compose -f redis.yaml up -d
```

We shall create the location service project and add the `gin` and `redis-go` dependencies.

Since our tech stack will be the same, we should execute the same initialization steps we executed in *Chapter 2, Running the First Application Using Compose*.

These are the initialization steps:

```
go mod init location_service
go get github.com/gin-gonic/gin
go get github.com/go-redis/redis/v8
```

After this, we shall create the `main.go` file, which contains the base of our application. We shall use the `gin` framework as we did before, as well as the Redis database; therefore, we should use the same helper methods, which we can get from GitHub: `https://github.com/PacktPublishing/A-Developer-s-Essential-Guide-to-Docker-Compose/blob/main/Chapter5/location-service/main.go`.

Since we have the project set, we can proceed with the application logic. An important part of our service is the location model. The location model will hold information such as the unique ID we give to that location, the longitude and latitude, the name of that location, as well as its description. Since we use a REST API, the location model will be marshaled to JSON and returned through the API calls.

The location model to be used in the code base is as follows:

```
type Location struct {
    Id          string  `json:"id"`
    Name        string  `json:"name"`
    Description string  `json:"description"`
    Longitude   float64 `json:"longitude"`
    Latitude    float64 `json:"latitude"`
}
```

This service is based on the concept of geolocation; thus, the proper Redis data structures need to be chosen. The location model can be represented in `hmset`. `hmset` makes it possible to fetch the object in a key-value manner. By using a prefix and the ID of the object (`location:id`), we can have multiple location objects. Also, thanks to the functionality of `hmset`, we can fetch the individual members of the objects.

Another important aspect of a location is distance. We would like to be able to retrieve locations in our database based on a location and up to a certain distance. For this purpose, Redis provides us with the Geohash technique. By using GeoAdd, we add locations to a sorted set. A hash is generated using the latitude and longitude and is used as a ranking by the various Geohash functions that can operate on a sorted set. This makes it feasible, for example, to find locations that are within a certain distance from a location or even calculate the distance between two locations stored on the sorted set. Based on these details, to store a location, we shall store it to a hash and make an entry to a sorted set using GeoAdd.

The function that persists the location is as follows:

```go
// Chapter5/location-service/main.go:177
func persistLocation(c context.Context, location Location)
error {
    hmset := client.HSet(c,
            fmt.Sprintf(locationIdFormat, location.Id), "Id",
            location.Id, "Name",
            location.Name, "Description",
            location.Description, "Longitude",
            location.Longitude, "Latitude",
            location.Latitude)

    if hmset.Err() != nil {
            return hmset.Err()
    }

    geoLoc := &redis.GeoLocation{Longitude: location.
Longitude, Latitude: location.Latitude, Name: location.Id}

    gadd := client.GeoAdd(c, "locations", geoLoc)

    if gadd.Err() != nil {
            return gadd.Err()
    }

    return nil
}
```

This is the method that retrieves the location from the hash:

```go
// Chapter5/location-service/main.go:153
func fetchLocation(c context.Context, id string) (*Location,
error) {
    hgetAll := client.HGetAll(c, fmt.Sprintf(locationIdFormat,
id))

    if err := hgetAll.Err(); err != nil {
        return nil, err
    }

    ires, err := hgetAll.Result()

    if err != nil {
        return nil, err
    }

    if l := len(ires); l == 0 {
        return nil, nil
    }

    latitude, _ := strconv.ParseFloat(ires["Latitude"], 64)
    longitude, _ := strconv.ParseFloat(ires["Longitude"], 64)

    location := Location{Id: ires["Id"], Name: ires["Name"],
Description: ires["Description"], Longitude: longitude,
Latitude: latitude}
    return &location, nil
}
```

Since we would like to retrieve existing locations based on distance and location, we shall use the Redis spatial functions. A method will be implemented, using the GEORADIUS method, on the set that we added elements to using GEOADD previously:

```go
// Chapter5/location-service/main.go:124
[...]
func nearByLocations(c context.Context, longitude float64,
latitude float64, unit string, distance float64) ([]
```

```
LocationNearMe, error) {
    var locationsNearMe []LocationNearMe = make([]
LocationNearMe, 0)
    query := &redis.GeoRadiusQuery{Unit: unit, WithDist: true,
Radius: distance, Sort: "ASC"}
    geoRadius := client.GeoRadius(c, "locations", longitude,
latitude, query)

    if err := geoRadius.Err(); err != nil {
        return nil, err
    }

    geoLocations, err := geoRadius.Result()

    if err != nil {
        return nil, err
    }

    for _, geoLocation := range geoLocations {
        if location, err := fetchLocation(c, geoLocation.
Name); err != nil {
            return nil, err
        } else {
            locationsNearMe = append(locationsNearMe,
LocationNearMe{
                Location: *location,
                Distance: geoLocation.Dist,
            })
        }
    }

    return locationsNearMe, nil
}
[...]
```

Based on the distance from the coordinates and the distance limit provided, the locations closest to the point will be returned in ascending order.

Since the core methods are implemented, we shall create the REST API using `gin`:

```go
// Chapter5/location-service/main.go:49
[...]
r.GET("/location/:id", func(c *gin.Context) {
    id := c.Params.ByName("id")

    if location, err := fetchLocation(c.Request.Context(),
id); err != nil {
        [...]
    } else if location == nil {
        [...]
    } else {
        [...]
    }

})

r.POST("/location", func(c *gin.Context) {
    var location Location

    [...]
    if err := persistLocation(c, location); err != nil {
        c.JSON(http.StatusInternalServerError,
gin.H{"location": location, "created": false, "message": err.
Error()})
        return
    }

    [...]
})

r.GET("/location/nearby", func(c *gin.Context) {
    [...]

    if locationsNearMe, err := nearByLocations(c, longitude,
latitude, unit, distance); err != nil {
```

```
            c.JSON(http.StatusInternalServerError,
gin.H{"message": err.Error()})
            return
    } else {
            c.JSON(http.StatusOK, gin.H{"locations":
locationsNearMe})
        }


})
```

We shall run the application and execute some requests:

```
$ go run main.go
```

And location using curl:

```
$ curl --location --request POST 'localhost:8080/location/' \
--header 'Content-Type: application/json' \
--data-raw '{
        "id":"0c2e2081-075d-443a-ac20-40bf3b320a6f",
        "name": "Liverpoll Street Station",
    "description": "Station for Tube and National Rail",
        "longitude": -0.082966,
    "latitude": 51.517336
}'
{"created":true,"location":{"id":"0c2e2081-075d-
443a-ac20-40bf3b320a6f","name":"Liverpoll Street
Station","description":"Station for Tube and National
Rail","Longitude":-0.082966,"Latitude":51.517336},"message":"Lo
cation Created Successfully"}
```

Find a location using the location ID:

```
$ curl --location --request GET 'localhost:8080/
location/0c2e2081-075d-443a-ac20-40bf3b320a6f'
{"location":{"id":"0c2e2081-075d-443a-ac20-
40bf3b320a6f","name":"Liverpoll Street
Station","description":"Station for Tube and National
Rail","Longitude":-0.082966,"Latitude":51.517336}}
```

Finally, we will retrieve the location near a point specified, limited by a certain distance. Because we would like to know the distance to each location, we need a new model. The model will contain the location itself as well as the distance calculated:

```
type LocationNearMe struct {
    Location Location `json:"location"`
    Distance float64 `json:"distance"`
}
```

Retrieve the locations near to a point using `curl`:

```
curl --location --request GET 'localhost:8080/location/
nearby?longitude=-0.0197&latitude=51.5055&distance=5&unit=km'
{"locations":[{"location":{"id":"0c2e2081-075d-
443a-ac20-40bf3b320a6f","name":"Liverpoll Street
Station","description":"Station for Tube and National
Rail","Longitude":-0.082966,"Latitude":51.517336},"dista
nce":4.5729}]}
```

To assist the preceding commands, a Postman collection can be found in the book's repository (https://github.com/PacktPublishing/A-Developer-s-Essential-Guide-to-Docker-Compose/blob/main/Chapter5/location-service/Location%20 Service.postman_collection.json).

We have created our first microservice in this section. We stored the locations and utilized the spatial capabilities of Redis to assist in searching for existing locations. The Task Manager should interact with the location service by using the REST interface provided. In the next section, we shall package the application in a Docker image and create the Compose application.

Adding a location service to Compose

We have implemented the service and have been able to add locations and execute spatial queries. The next step is to package the application using Docker and run it through Compose.

The first step is to create the Dockerfile. The same steps we followed in the previous chapter also apply here:

1. Adding a Dockerfile
2. Building the image using Compose

This is the Dockerfile for the location service:

```
# syntax=docker/dockerfile:1

FROM golang:1.17-alpine

RUN apk add curl

WORKDIR /app

COPY go.mod ./
COPY go.sum ./

RUN go mod download

COPY *.go ./

RUN go build -o /location_service

EXPOSE 8080

CMD [ "/location_service" ]
```

The Dockerfile is in place, and we can now proceed to run it through Compose. Now, to test the application, we need Redis and the image of the application to be built.

Our docker-compose.yaml, at this stage, should look like this:

```
services:
  location-service:
    build:
      context: ./location-service
    image: location-service:0.1
    environment:
      - REDIS_HOST=redis:6379
    depends_on:
      - redis
    healthcheck:
```

```
      test: ["CMD", "curl", "-f", "http://localhost:8080/ping"]
      interval: 10s
      timeout: 5s
      retries: 5
      start_period: 5s
  redis:
    image: redis
```

Note that the `ports` section is missing. This is because the location service is going to be a private one. It shall be accessed only by other containers hosted on Compose.

An image called `location-service` will be built once we spin up the Compose application. The Redis image should work as it is.

We managed to package the `location-service` microservice using Docker and run it through Compose. Since we will introduce the Task Manager, we need to revise the networks that the Compose application will provision.

Adding a network for the location microservice

We can now specify the network on which the application shall run instead of using the default network, as we did previously. The network shall be named `geolocation-network`, and we also need a network for Redis. We shall add those networks to Compose:

```
services:
  location-service:
[...]
    networks:
      - location-network
      - redis-network
[...]
  redis:
    image: redis
    networks:
      - redis-network
networks:
  location-network:
  redis-network:
```

Redis does not expose any port locally; the geolocation service is able to access the service only because it has `redis-network` included in the `networks` section. `redis-network` is a familiar name. It is the same network name we used in *Chapter 3, Network and Volumes Fundamentals*. Since our microservice is up and running on a dedicated network, we can now proceed with integrating it with the Task Manager application.

Executing requests to the location microservice

Previously, we successfully ran the recently introduced location microservice using Compose. However, the application will be unusable if we do not use it along with the Task Manager. By integrating the Task Manager with the location service, the user should be able to specify a location when creating a task. If the user also retrieves one of the existing tasks, locations near the task's location shall be presented.

The Task Manager would have to communicate with the location service. For this reason, we shall create a service inside the Task Manager that will issue requests to the location microservice. The same models we used on the location service will also be used for this module.

The location service module in the Task Manager application can be found on GitHub: `https://github.com/PacktPublishing/A-Developer-s-Essential-Guide-to-Docker-Compose/blob/main/Chapter5/task-manager/location/location_service.go`.

Since the Task Manager will have various feature additions in this chapter, it makes sense to refactor the code. The models should change and include the location models we defined previously:

```
type Task struct {
    Id            string             `json:"id"`
    Name          string             `json:"name"`
    Description   string             `json:"description"`
    Timestamp     int64              `json:"timestamp"`
    Location      *location.Location `json:"location"`
}
```

Also, we shall separate the persistence methods from the controller method and move them to another file, `task_service.go` (`https://github.com/PacktPublishing/A-Developer-s-Essential-Guide-to-Docker-Compose/blob/main/Chapter5/task-manager/task/task_service.go`).

If we take a close look at the logic, the user is free to add just a task without specifying the location. If the location is specified and the ID already exists, it will pass through without any persistence to the location service. If the location is specified and does not exist, then we can specify the entire location payload and persist it. By specifying only an existing location ID, we do not have to specify the entire payload. Since we modularized the Task Manager code base, we can proceed to adapt the `http` controllers (`https://github.com/PacktPublishing/A-Developer-s-Essential-Guide-to-Docker-Compose/blob/main/Chapter5/task-manager/main.go`).

Now that the Task Manager is adapted, we can update the Compose application and add the Task Manager interacting with location-service.

There are some requirements before doing so:

- The Task Manager requires Redis and location-service to be up and running.

- The Task Manager needs to have access to the networks of the preceding services.

- The Task Manager is an entry point; thus, we shall bind the port to host.

The Docker image shall be built the same way we did in *Chapter 2, Running the First Application Using Compose*, but we do need to add the extra source code created previously:

```
# syntax=docker/dockerfile:1
FROM golang:1.17-alpine
RUN apk add curl
WORKDIR /app
RUN mkdir location
RUN mkdir task
RUN mkdir stream
COPY go.mod ./
COPY go.sum ./
RUN go mod download
COPY *.go ./
COPY location/*.go ./location
COPY task/*.go ./task
COPY stream/*.go ./stream
RUN go build -o /task_manager
EXPOSE 8080
CMD [ "/task_manager" ]
```

We can now see the Compose file, including the Task Manager and the location service:

```
// Chapter5/task-manager/docker-compose.yaml:19
[...]
  task-manager:
    build:
      context: ./task-manager/
    image: task-manager:0.1
    ports:
```

```
      - 8080:8080
    environment:
      - REDIS_HOST=redis:6379
      - LOCATION_HOST=http://location-service:8080
    depends_on:
      - redis
      - location-service
    networks:
      - location-network
      - redis-network
    healthcheck:
      test: ["CMD", "curl", "-f", "http://localhost:8080/ping"]
      interval: 10s
      timeout: 5s
      retries: 5
      start_period: 5s
 [...]
```

The Task Manager was created and integrated with the location service. The location service remained an internal microservice without the need to expose it. The Task Manager established communication through a REST endpoint. In the next section, we shall evaluate accessing the application using a message-based form of communication.

Streaming task events

We have been successful previously in running the new microservice using Compose. However, we would like to know how many times a location has been visited or how many tasks have been created over time.

This is a data-driven task. We want to capture and stream information about our application. Redis provides us with streams. By using streams, our application can stream data that can later be processed by another application and create the analytics of our choice.

This will be possible with a simple adaptation to our code. Once a task is added, a message shall be published to a Redis stream.

We will add a service to the Task Manager that will be able to stream events. For now, only when adding a task will a message be sent.

The following code base is the implementation of the TaskStream service, which will be responsible for sending messages on task creation:

```
// Chapter5/task-manager/task/task-service.go:14
[...]
type TaskMessage struct {
    taskId      string
    location_id string
    timestamp   int64
}

func CreateTaskMessage(taskId string, location *location.
Location, timestamp int64) TaskMessage {
    taskMessage := TaskMessage{
            taskId:    taskId,
            timestamp: timestamp,
    }

    if location != nil {
            taskMessage.location_id = location.Id
    }

    return taskMessage
}

func (ts *TaskMessage) toXValues() map[string]interface{} {
    return map[string]interface{}{"task_id": ts.taskId,
"timestamp": ts.timestamp, "location_id": ts.location_id}
}

func (ts *TaskStream) Publish(c context.Context, message
TaskMessage) error {

    cmd := ts.Client.XAdd(c, &redis.XAddArgs{
            Stream: "task-stream",
            ID:     "*",
            Values: message.toXValues(),
```

```
    })

    if _, err := cmd.Result(); err != nil {
        return err
    }

    return nil
}
```

Since we have the message-sending functionality implemented, we will change the `PersistTask` method in order to send an update once a task is created:

```
// Chapter5/task-manager/task/task-service.go:28
[...]
func (ts *TaskService) PersistTask(c context.Context, task
Task) error {

    values := []interface{}{"Id", task.Id, "Name", task.Name,
"Description", task.Description, "Timestamp", task.Timestamp}

    if task.Location != nil {
        if err := ts.LocationService.AddLocation(task.
Location); err != nil {
            return err
        }
        values = append(values, "location", task.Location.Id)
    }

    hmset := ts.Client.HSet(c, fmt.Sprintf("task:%s", task.
Id), values)

    if hmset.Err() != nil {
        return hmset.Err()
    }

    z := redis.Z{Score: float64(task.Timestamp), Member: task.
Id}
    zadd := ts.Client.ZAdd(c, "tasks", &z)
```

```
    if zadd.Err() != nil {
        return hmset.Err()
    }

    mes := stream.CreateTaskMessage(task.Id, task.Location,
task.Timestamp)

    return ts.TaskStream.Publish(c, mes)
}
[...]
```

So far, we have enhanced our application to send events on task insertions. In the next section, we shall proceed with consuming those messages.

Adding a task events processing microservice

In the previous section, we produced events regarding our Task Manager application. This enables us to add an application that shall be message-driven. For now, the events service will consume the data from a Redis stream and print data on the console.

Our code base will be lean and require only the Redis client.

Let's add the code that will consume the events:

```
[...]
client. XGroupCreateMkStream (ctx, stream, consumerGroup, "0").
Result()

for {
    entries, err := client.XReadGroup(ctx,
        &redis.XReadGroupArgs{
            Group:    consumerGroup,
            Consumer: consumer,
            Streams:  []string{stream, ">"},
            Count:    1,
            Block:    0,
            NoAck:    false,
        },
    ).Result()
```

```
    for i := 0; i < len(entries[0].Messages); i++ {
        messageID := entries[0].Messages[i].ID
        values := entries[0].Messages[i].Values

        taskId := values["task_id"]
        timestamp := values["timestamp"]
        locationId := values["location_id"]

        log.Printf("Received %v %v %v", taskId, timestamp,
locationId)

        client.XAck(ctx, stream, consumerGroup, messageID)
    }
}
[...]
```

Let's create the Dockerfile for it:

```
# syntax=docker/dockerfile:1

FROM golang:1.17-alpine

RUN apk add curl

WORKDIR /app

COPY go.mod ./
COPY go.sum ./

RUN go mod download

COPY *.go ./

RUN go build -o /events_service

CMD [ "/events_service" ]
```

Then, we should add a task service to Compose:

```
services:
  location-service:
    [...]
  task-manager:
    [...]
  event-service:
    build:
      context: ./events-service/
    image: event-service:0.1
    environment:
      - REDIS_HOST=redis:6379
    depends_on:
      - redis
    networks:
      - redis-network
networks:
  location-network:
  redis-network:
```

It is obvious that the new service does not need to have as many settings on Compose as the REST-based service. Being stream-based, it only needs the connection to the Redis stream.

Since all the service's configurations are in place, we can run the application and observe task events getting streamed to the events microservice:

```
$ docker compose up
...
chapter5-event-service-1      | 2022/05/08 09:03:38 Received
8b171ce0-6f7b-4c22-aa6f-8b110c19f83a 1645275972000 0c2e2081-
075d-443a-ac20-40bf3b320a6f
...
```

We have been successful in listening to events when a task is created, and incorporated that code into our existing Compose application.

Summary

In this chapter, we created two microservices that will integrate with the Task Manager. The microservices had a different nature; one used REST-based communication, and the other was message-driven. Regardless of the differences, by using Compose, it was possible to orchestrate those microservices and isolate them. As expected, monitoring plays a crucial part in the service we created. By monitoring properly, we can ensure availability and smooth usage for the end user.

The next chapter will be focused on monitoring and how to achieve it by using Prometheus.

6

Monitoring Services with Prometheus

In the previous chapter, we managed to create multiple services for the Task Manager application and orchestrated them using Compose. By transforming the Task Manager application into a microservice-based application, we made code adjustments and facilitated the communication between multiple microservices.

The communication was either REST-based or message-based. By establishing the communication between the microservices, we managed to highlight the network features of Compose and made the distinction between services that were operating privately and services that were publicly accessible.

This chapter will be focused on the monitoring of the services through Compose. When it comes to microservices, we need to monitor our services and effectively troubleshoot any issues that might arise. We did have some monitoring commands through Compose in *Chapter 4*, *Executing Docker Compose Commands*; however, we would like to gain more insights. Our goal is to have monitoring and metrics for our applications. Those metrics should be stored in a database, and we should be able to retrieve them using queries. For this reason, Prometheus would be the tool of our choice in this chapter.

In this chapter, the following topics will be covered:

- What is Prometheus?
- Adding an endpoint for Prometheus
- Configuring Prometheus to parse metrics
- Adding Prometheus to the Compose network
- Your first metrics query

What is Prometheus?

Prometheus is a popular open source monitoring solution with a wide range of capabilities including event monitoring and alerting. Prometheus follows an HTTP `pull` model. The applications expose their metrics through an HTTP endpoint. Prometheus has the ability to parse the exposed metrics. By configuring certain applications as targets on Prometheus, Prometheus will proceed with parsing the exposed metrics. Additionally, for services where exposing an HTTP endpoint is not feasible, Prometheus offers **Pushgateway**, an intermediary service where other services can push their data.

Once retrieved, the data is recorded inside a real-time series database, which is part of Prometheus. This makes it possible to have flexible queries and real-time alerting.

Here are some of the features that Prometheus offers:

- A data model for time series metrics
- A query language to execute queries upon the retrieved time series data
- Storing metrics in one single autonomous server
- A `pull` model over HTTP for collecting data
- Pushing metrics through an intermediary gateway, for non-HTTP applications
- Discovery or configuration of targets exposing metrics
- Dashboard capabilities

In this chapter, we will adapt our application and help them to export data to Prometheus. We will take an extensive look at the features that Prometheus provides and create a dashboard for our applications.

Adding an endpoint for Prometheus

Let's see how to add an endpoint to our existing Compose endpoints. The code base that will be used in this chapter will be the code base that we created in *Chapter 5, Connecting Microservices*. We should update the code base and add the endpoints that will enable Prometheus to scrap metrics from our applications.

Adding the metrics endpoint to the Task Manager

Adding Prometheus to an HTTP-based Go application is streamlined. By following the instructions online (`https://prometheus.io/docs/guides/go-application`), we can find the following `go get` commands that will download the necessary libraries to use with our Go application:

```
$ go get github.com/prometheus/client_golang/prometheus
$ go get github.com/prometheus/client_golang/prometheus/
promauto
```

```
$ go get github.com/prometheus/client_golang/prometheus/
promhttp
```

The default way to add a Prometheus endpoint is by using the Go http server:

```
func main() {
        http.Handle("/metrics", promhttp.Handler())
        http.ListenAndServe(":2112", nil)
}
```

Task Manager is based on the go framework and the gin framework. This will require a minor workaround since gin offers a wrapper for the http handlers:

```
import (
...
"github.com/prometheus/client_golang/prometheus/promhttp"
...
)
...
// Metrics Endpoint
r.GET("/metrics", gin.WrapH(promhttp.Handler()))
...
```

Now if we run the application, we should be able to access the metrics:

```
$ curl localhost:8080/metrics
# HELP go_gc_cycles_automatic_gc_cycles_total Count of
completed GC cycles generated by the Go runtime.
# TYPE go_gc_cycles_automatic_gc_cycles_total counter
go_gc_cycles_automatic_gc_cycles_total 0
# HELP go_gc_cycles_forced_gc_cycles_total Count of completed
GC cycles forced by the application.
# TYPE go_gc_cycles_forced_gc_cycles_total counter
go_gc_cycles_forced_gc_cycles_total 0
# HELP go_gc_cycles_total_gc_cycles_total Count of all
completed GC cycles.
# TYPE go_gc_cycles_total_gc_cycles_total counter
go_gc_cycles_total_gc_cycles_total 0
...
```

We have just retrieved the metrics for our application, the metrics that have been collected since its start. We can observe various metrics such as memory, garbage collection, and more.

Adding the metrics endpoint to the location service

Just like the Task Manager, the location service is also based on `go gin`; therefore, exactly the same steps should be followed. The results should be the same, and you can check them on GitHub (`https://github.com/PacktPublishing/A-Developer-s-Essential-Guide-to-Docker-Compose/tree/main/Chapter6/location-service`).

Exporting metrics from the Event Service

The Event Service is a microservice but an event-driven one. The events are received through a Redis stream; therefore there is no need for an HTTP endpoint to be set up. In order to enable monitoring using Prometheus, we can either add an HTTP server on the application just for this purpose or we can enable the use of the Pushgateway service mentioned earlier.

We would like to track how fast the messages that are received are processed. Therefore, every time a new message is received and processed, we should submit those points to Prometheus using the Pushgateway service.

For our use case, we will use a gauge. Based on the documentation, a gauge is a metric that represents a single numerical value that can arbitrarily go up and down (`https://prometheus.io/docs/concepts/metric_types/#gauge`).

For now, we will assume that the Pushgateway service is already present and will accept the presented metrics.

The following method will be the metrics to push:

```
...
processingTime = prometheus.NewGauge(prometheus.GaugeOpts{
    Name: "task_event_process_duration",
    Help: "Time it took to complete a task",
})

processedCounter = prometheus.NewCounterVec(
    prometheus.CounterOpts{
        Name: "task_event_processing_total",
        Help: "How many tasks have been processed",
    },
    []string{"task"},
```

```
).WithLabelValues("task")
...
```

In the main processing method, we will call the method once the message has been processed:

```
...
start := time.Now()
log.Printf("Received %v %v %v", taskId, timestamp, locationId)

client.XAck(ctx, stream, consumerGroup, messageID)
elapsed := time.Since(start)

processedCounter.Add(1)

millis := float64(elapsed.Milliseconds())
processingTime.Set(millis)

pushProcessingDurationToPrometheus(processingTime)
pushProcessingCount(processedCounter)
...
```

And last but not least, we will add the methods that push the data toward the gateway:

```
...
func pushProcessingDurationToPrometheus(processingTime
prometheus.Gauge) {
    if err := push.New(getStrEnv("PUSH_GATEWAY", "http://
localhost:9091"), "task_event_process_duration").
        Collector(processingTime).
        Grouping("db", "event-service").
        Push(); err != nil {
        fmt.Println("Could not push completion time to
Pushgateway:", err)
    }
}

func pushProcessingCount(processedCounter prometheus.Counter) {
    if err := push.New(getStrEnv("PUSH_GATEWAY", "http://
localhost:9091"), "task_event_processing_total").
```

```
        Collector(processedCounter).
        Grouping("db", "event-service").
        Push(); err != nil {
        fmt.Println("Could not push tasks processed to
Pushgateway:", err)
    }
}
...
```

So, we have managed to apply the changes to our code base and made it possible for our applications to expose their metrics. The next step is to set up a Prometheus instance in order to parse the exposed data. So, let's proceed and see how a Prometheus instance will interact with one of our services.

Configuring Prometheus to parse metrics

As mentioned earlier, Prometheus operates using a `pull`-based model. This means that the data is exposed through an HTTP endpoint and Prometheus should parse it, provided it has been configured properly to access it.

In order for Prometheus to parse from an endpoint, a job configuration needs to be applied. Job configurations for Prometheus are in YAML format.

In our scenario, we would like to parse metrics with an interval of one minute, from the Task Manager.

The configuration should look like this:

```
scrape_configs:
  - job_name: 'task-manager'
    scrape_interval: 1m
    metrics_path: '/metrics'
    static_configs:
      - targets: ['host.docker.internal:8080']
```

Provided the Task Manager application is running and the YAML file is ready, we can create the Compose YAML file:

```
services:
  prometheus:
    image: prom/prometheus
    ports:
      - 9090:9090
```

```
        volumes:
            - ./prometheus.yaml:/etc/prometheus/prometheus.yml
```

By running the following code, Prometheus will pick the specified configuration:

```
$ docker compose -f ./prometheusold.yaml up
```

After waiting a few minutes, we can navigate to the Prometheus URL. We should see the results of the scraping on our screen:

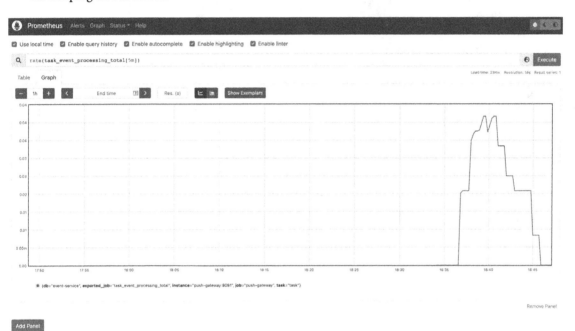

Figure 6.1 – Metrics plotted on the Prometheus dashboard

Also, in the configuration tab, we can see the configuration we applied:

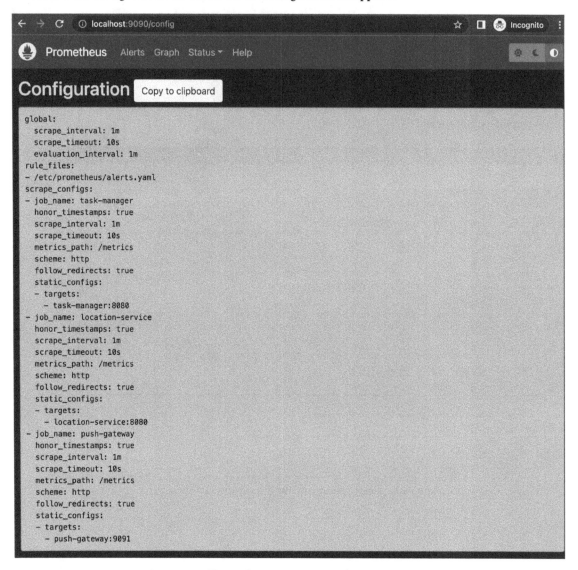

Figure 6.2 – The configuration retrieved from Prometheus

We managed to monitor Task Manager through Prometheus by making it possible for Prometheus to scrape some metrics. In the next section, we will apply Prometheus to all of our services.

Adding Prometheus to the Compose network

We have set Prometheus up so that it is running successfully and also managed to monitor the Task Manager application. Since we added Prometheus to all of our applications, we should add it to the Compose file and parse the services.

Since Prometheus is interacting with the existing services, this is not an operation that should place from an external network. We should add a network that Prometheus will be able to operate. We will name it `monitoring-network`:

```
networks:
  location-network:
  redis-network:
  monitoring-network:
```

Then, we should create the Prometheus configuration file. So far, we have introduced three services:

- `task-manager`
- `location-service`
- `events-service`

Here, `task-manager` and `location-service` are on port `8080`. The configuration should look like this:

```
scrape_configs:
  - job_name: 'task-manager'
    scrape_interval: 1m
    metrics_path: '/metrics'
    static_configs:
      - targets: ['task-manager:8080']
  - job_name: 'location-service'
    scrape_interval: 1m
    metrics_path: '/metrics'
    static_configs:
      - targets: ['location-service:8080']
```

As we can see, we take advantage of Compose's internal DNS. Let's put the Prometheus configuration and add other services to the same network:

```
prometheus:
  image: prom/prometheus
```

```
    ports:
      - 9090:9090
    volumes:
      - ./prometheus.yaml:/etc/prometheus/prometheus.yml
    networks:
      - monitoring-network
```

We should attach each service to the Docker network.

Note that both `task-manager` and `location-service` will have their networks altered:

```
    ...
    networks:
      - location-network
      - redis-network
      - monitoring-network
    ...
```

By running the Compose application, Prometheus will be able to operate and parse the metrics.

We can see the results by accessing the running Prometheus instance at `localhost:9090`.

Pushing metrics to Prometheus

By running the Compose application, Prometheus will be able to operate and parse the metrics. However, there is a need to push the metrics instead of exposing them. Our analytics services will benefit from this purpose. To do so, we should enable the Prometheus gateway.

Let's add the gateway Compose configuration:

```
// Chapter6/docker-compose.yaml:68
  push-gateway:
    image: prom/pushgateway
    networks:
      - monitoring-network
```

By having the Pushgateway service available, the Event Service can push metrics to Prometheus. By pushing the metrics to Pushgateway, the gateway will expose them through an endpoint. Then, Prometheus should be configured to pull the messages from the gateway:

```
// Chapter6/prometheus.yaml
  - job_name: 'push-gateway'
```

```
    scrape_interval: 1m
    metrics_path: '/metrics'
    static_configs:
      - targets: ['push-gateway:9091']
```

Based on the preceding steps, we can use the code we changed in the event-service application. Now, all of our services are monitored and push data to Kubernetes. As expected, we can run the entire application using compose up:

```
$ docker compose build
$ docker compose up -d
```

Now we can proceed to the next section and make use of what we have built.

Creating your first metrics query

We managed to submit our metrics to Prometheus, so now we can execute queries upon metrics. By having metric data, we can now create dashboards in Prometheus.

Let's suppose we have lots of tasks created over time and we want to monitor them. To simulate the task creation, we will use the following script:

```
while true; do
curl --location --request POST 'localhost:8080/task/' \
--header 'Content-Type: application/json' \
--data-raw '{
    "id": "'$(date +%s%N)'",
    "name": "A task",
    "description": "A task that need to be executed at the
timestamp specified",
    "timestamp": 1645275972000,
    "location": {
        "id": "1c2e2081-075d-443a-ac20-40bf3b320a6f",
        "name": "Liverpoll Street Station",
        "description": "Station for Tube and National Rail",
        "longitude": -0.081966,
        "latitude": 51.517336
    }
}'
```

```
sleep 1
done
```

The preceding script runs every second and creates a different task with a name generated by the date command in bash.

Moving forward, we will create a PromQL query to check the rate of tasks created over time.

The query in PromQL should be like this:

```
rate(task_event_processing_total[5m])
```

The dashboard will display the following result:

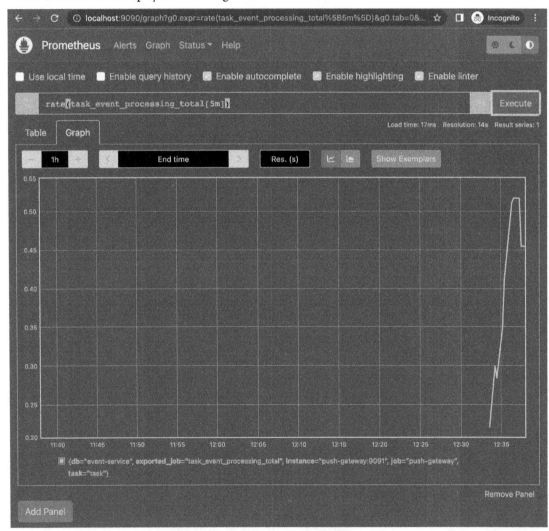

Figure 6.3 – The metric query plotted on Prometheus

Now we can also exit the script we run previously using *Ctrl + C*.

By submitting our metrics, we can plot queries and track incidents. This can help us to monitor our application. However, it requires our constant attention. We need to have a mechanism in place that will notify us of certain events that occur. Therefore, we should utilize Prometheus' alerts capabilities.

Adding an alert

Another thing Prometheus provides us with is the capability of alerts. In our case, if the rate of tasks over 5 minutes is bigger than 0.2 for 1 minute, an alert should be raised:

```
// Chapter6/alerts.yaml
groups:
- name: task-manager
  rules:
  - alert: too-many-tasks
    expr: rate(task_event_processing_total[5m]) > 0.2
    for: 1m
    annotations:
        summary: Too many tasks
```

We should mount this file to Prometheus; therefore, add it to the configuration of the prometheus. yaml file:

```
// Chapter6/prometheus.yaml:17
...
rule_files:
  - '/etc/prometheus/alerts.yaml'
```

After some tasks are continuously created, we get the expected results.

Initially, we see that no alerts have been raised:

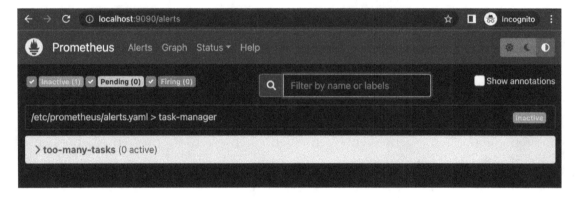

Figure 6.4 – The Prometheus alert screen

As we proceed and more alerts are added, we get to the pending state:

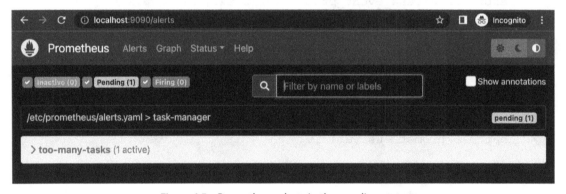

Figure 6.5 – Prometheus alerts in the pending state

Eventually, we should end up in the alert state:

Figure 6.6 – The Prometheus alerts triggered

We have managed to create alerts based on the queries we plotted earlier. By extracting those queries, we added them to a configuration file and mounted them to the Prometheus container through the Compose configuration.

Summary

In this chapter, we set up Prometheus on Compose and made it feasible to collect and receive metrics and raise alerts. By doing so, the multi-container application we created using Compose could submit metrics. This enabled us to monitor our application and get alerted if something was wrong. All of this has been made possible by changing our Compose installation and simply adding some extra configuration files using volumes on Compose. One issue with the preceding approach is the lack of flexibility. We need to have all services in one file running regardless if we only want to run individual components.

In the next chapter, we will evaluate how we can modularize our Compose application and run it on multiple files.

7

Combining Compose Files

So far, we have run our multi-container application in a monolithic way, where the application is run by specifying a Compose file that contains application containers, containers for databases such as Redis, and applications for monitoring purposes such as Prometheus. This will serve us well in the beginning; however, always running the application with all the dependencies available might bring us issues. Running a full-fledged application can consume many resources, it can be harder to troubleshoot issues, and it can prevent you from being focused on a certain component that requires your attention. There could be scenarios where you might want to use only one component and avoid interacting with or initiating other components. Also, there might be cases where you don't want to have monitoring enabled or any other stack that assists your application but is not directly related to the scope of your application.

Compose provides us with the option to split an application into multiple files and run the entire application by combining those files. This will give us the ability to run our application in a more modular way. We should be able to run certain parts of the application and completely ignore an entire stack.

In this chapter, we will proceed with splitting our application into multiple files and running them in a modular way.

In this chapter, we will cover the following topics:

- Splitting Compose files
- Combining Compose files
- Selecting the Compose files to run
- Creating different environments
- Combining multiple Compose files into one

Technical requirements

The code for this book is hosted on the GitHub repository at `https://github.com/PacktPublishing/A-Developer-s-Essential-Guide-to-Docker-Compose`. In case of an update to the code, it will be updated on the GitHub repository.

Splitting Compose files

Throughout the Task Manager application development, we started using one simple Go application backed by a Redis database. Onward, we enhanced the functionality of our main application by adding two extra microservices. Since we ended up with a full-functional microservice-based application, we reckoned that more monitoring was needed; therefore, we added Prometheus and the Pushgateway to facilitate proper monitoring of our applications. Each step is required to incorporate the services into the Docker Compose file.

If we properly examine each step, we could identify components that are shared between applications and need to be available regardless of the applications we want to run. Those are core components that we should share with other services; thus, they can be logically grouped into a Compose file. More specifically, the networks and the database are a part of our core Compose project, which is essential for our application to run.

In our application, we can also identify that certain services can have their own compose file. For example, the location service can run standalone provided it has a service database available. The same applies to the event service.

The Prometheus stack is also something that can run separately since it is not directly related to the goal of our applications server. However, they are essential for running our applications properly.

So, let's proceed to create the base of our compose application.

Task Manager base

The `task-manager` base would be from the networks and the Redis database. The definition of the Docker networks is essential since they are the backbone of our application's connectivity. The database where we store data is also essential since all of our applications need to be backed by a database assisting them to store the data.

Therefore, at the root of our project, we will create the `base-compose.yaml` file:

```
// Chapter7/base-compose.yaml
services:
  redis:
    image: redis
    networks:
```

```
        - redis-network
  networks:
    location-network:
    redis-network:
    monitoring-network:
```

If we spin up the base service, we should see the Redis database up:

```
// Chapter7/base-compose.yaml
$ docker compose -f base-compose.yaml up -d
$ docker compose -f base-compose.yaml ps
NAME                 COMMAND                SERVICE
STATUS               PORTS
chapter7-redis-1     "docker-entrypoint.s…"  redis
running              6379/tcp
$ docker compose -f base-compose.yaml down
```

The base is ready. It provides us with a Redis database and the essential network configurations. Next, we will proceed with the location service.

Location service

The location service is the first Go-based service to have a Compose file dedicated only to run this service. We will extract the compose configuration we had previously and will use the components from the base-compose.yaml file.

The compose file for the location service is detailed as follows:

```
// Chapter7/location-service/docker-compose.yaml
services:
  location-service:
    build:
      context: location-service
    image: location-service:0.1
    environment:
      - REDIS_HOST=redis:6379
    depends_on:
      - redis
    networks:
      - location-network
```

```
      - redis-network
      - monitoring-network
    healthcheck:
      test: ["CMD", "curl", "-f", "http://localhost:8080/ping"]
      interval: 10s
      timeout: 5s
      retries: 5
      start_period: 5s
```

As you can see, extracting the Compose configuration for the location service is streamlined, and all we had to do was to copy the service configuration. However, we do see a small adaption. Instead of context: ., our code base is switched to context: location-service. The reason for this has to do with base-compose.yaml and location-service/docker-compose.yaml being located on different files. The location paths on Compose are absolute and follow the location of the first file specified. For this reason, we will use the context path from compose. The next service will be the event service.

Event service

The event service will also have no changes and will be moved to a separate file:

```
// Chapter7/event-service/docker-compose.yaml
services:
  event-service:
    build:
      context: event-service
    image: event-service:0.1
    environment:
      - REDIS_HOST=redis:6379
      - PUSH_GATEWAY=push-gateway:9091
    depends_on:
      - redis
    networks:
      - redis-network
      - monitoring-network
```

By having the first two services split up, we can proceed to the main service, the task-manager service.

Task Manager

We proceed to the `task-manager` service last, mainly because it is a service that interacts with the other two services.

By extracting the content to another file, the following would be the dedicated compose file:

```
// Chapter7/task-manager/docker-compose.yaml
services:
  task-manager:
    build:
      context: task-manager
    image: task-manager:0.1
    ports:
      - 8080:8080
    environment:
      - REDIS_HOST=redis:6379
      - LOCATION_HOST=http://location-service:8080
    depends_on:
      - redis
      - location-service
    networks:
      - location-network
      - redis-network
      - monitoring-network
    healthcheck:
      test: ["CMD", "curl", "-f", "http://localhost:8080/ping"]
      interval: 10s
      timeout: 5s
      retries: 5
      start_period: 5s
```

With Task Manager setup, we did split the Go-based services. Now, we can proceed with splitting up the Prometheus components.

Prometheus

Prometheus is targeted toward monitoring; therefore, it's going to be on a separate file, and whether it is going to be used among the other files will be up to the user. By not including Prometheus when running the `task-manager` application, the monitoring features will not be there. However, there would be fewer dependencies and fewer resources will be needed.

The Prometheus compose file will require the Prometheus server and the `push-gateway` included. This makes sense if we take into consideration that they are both monitoring solutions.

In the following compose file, we have the configuration for Prometheus and `push-gateway`:

```
// Chapter7/monitoring/docker-compose.yaml
services:
  prometheus:
    image: prom/prometheus
    ports:
      - 9090:9090
    volumes:
      - ./monitoring/prometheus.yaml:/etc/prometheus/
prometheus.yml
      - ./monitoring/alerts.yml:/etc/prometheus/alerts.yaml
    networks:
      - monitoring-network
    depends_on:
      - task-manager
  push-gateway:
    image: prom/pushgateway
    networks:
      - monitoring-network
```

By extracting the compose Prometheus components to another Compose file, we have modularized the `task-manager` application and can proceed with combining the files we created previously. Now, we should be able to use our application and have exactly the same behavior.

Combining Compose files

Now that we have split the `task-manager` application, we should be able to run and have the same functionality we had in *Chapter 6, Monitoring Services with Prometheus*. We should be able to store the tasks by executing requests to the Task Manager combined with a location.

Compose offers the option to combine multiple files.

Let's run the application and all the services needed together:

```
docker compose -f base-compose.yaml -f monitoring/docker-
compose.yaml -f event-service/docker-compose.yaml -f location-
service/docker-compose.yaml -f task-manager/docker-compose.yaml
up
 Network chapter7_location-network     Created     0.0s
 Network chapter7_redis-network        Created     0.0s
 Network chapter7_monitoring-network   Created     0.0s
 Container chapter7-redis-1            Created     0.0s
 Container chapter7-push-gateway-1     Created     0.0s
 Container chapter7-location-service-1 Created     0.0s
 Container chapter7-event-service1     Created     0.0s
 Container chapter7-task-manager 1     Created     0.0s
 Container chapter7-prometheus-1       Created     0.0s
```

An observation is that the prefix starts with chapter7. This has to do with the base compose file, which is the first on the list. base-compose.yaml is on the root of Chapter7; therefore, the relative path is set to Chapter7. This works well with the configuration we have set in the monitoring project.

So far, we have been successful in splitting our original application into parts and also running them all together. The qualities of the application have been the same, and the modularization of our application makes development easier. However, we are still not flexible enough. We still need to run the application including all of the files, and we are not able to select the application we want to focus on individually. In the next section, we will see how Compose can assist us in making the application modular.

Selecting the Compose files to run

In the previous section, one of the issues we stumbled upon is the fact that we run the application's Compose file altogether. However, modularization is in place since we split the compose file into multiple parts. Thus, the next step would be to run debug and test different modules of the application individually.

Using Hoverfly

Since our applications depend on each other, the only viable option is to run the applications together until we find an alternative. For development and testing purposes, we can mock some of the services that introduce dependencies and still be able to run our application locally.

For this purpose, Hoverfly (https://hoverfly.io/) can be of significant help. Hoverfly can intercept traffic and mock requests and responses.

We will spin up a Hoverfly instance with a capture mode in Compose:

```
services:
  hoverfly:
    image: spectolabs/hoverfly
    ports:
      - :8888
    networks:
      - location-network
      - monitoring-network
    entrypoint: ["hoverfly","-capture","-listen-on-
host","0.0.0.0"]
```

By having Hoverfly enabled, we will use it in order to intercept traffic and then use it to replay the traffic as we test our application.

Extending services

In order to have a modified version of the existing service without duplicating the content, Compose provides us with the feature of **extending services**.

By extending services, we import the existing compose file and make alterations to the components of interest.

Let's take a look at the following example:

```
services:
  db:
    extends:
      file: databases.yml
      service: postgresql
    environment:
      - AUTOVACUUM=true
```

Here, we created another Compose file that extends the postgresql service defined in the databases.yml Compose file and added an extra environment variable.

Capturing traffic with Hoverfly

Two services create HTTP traffic:

- `task-manager` toward `location-service`
- `event-service` toward `push-gateway`

In both cases, since the default `http` Go client is used, this makes it easy to set up using Hoverfly as a proxy through an `env` variable.

In this case, we will extend the `task-manager` and `event-service` services and enable the `http` proxy.

The following is the `event-service` adaptation:

```
services:
  event-service:
    extends:
      file: ./event-service/docker-compose.yaml
      service: event-service
    environment:
      - HTTP_PROXY=hoverfly:8500
    depends_on:
      - hoverfly
```

The following is the `task-manager` adaptation:

```
// Chapter7/task-manager/capture-traffic-docker-compose.yaml
services:
  task-manager:
    extends:
      file: ./task-manager/docker-compose.yaml
      service: task-manager
    environment:
      - HTTP_PROXY=hoverfly:8500
    depends_on:
      - hoverfly 0
```

By using this configuration, we can capture the traffic exchanged. Then, we can run the stack together and check whether Hoverfly has captured anything:

```
docker compose -f base-compose.yaml -f monitoring/docker-
compose.yaml -f event-service/capture-traffic-docker-compose.
yaml -f location-service/docker-compose.yaml -f task-manager/
capture-traffic-docker-compose.yaml -f hoverfly/docker-compose.
yaml up
```

After creating some tasks, we can navigate to Hoverfly and check for requests that have been captured:

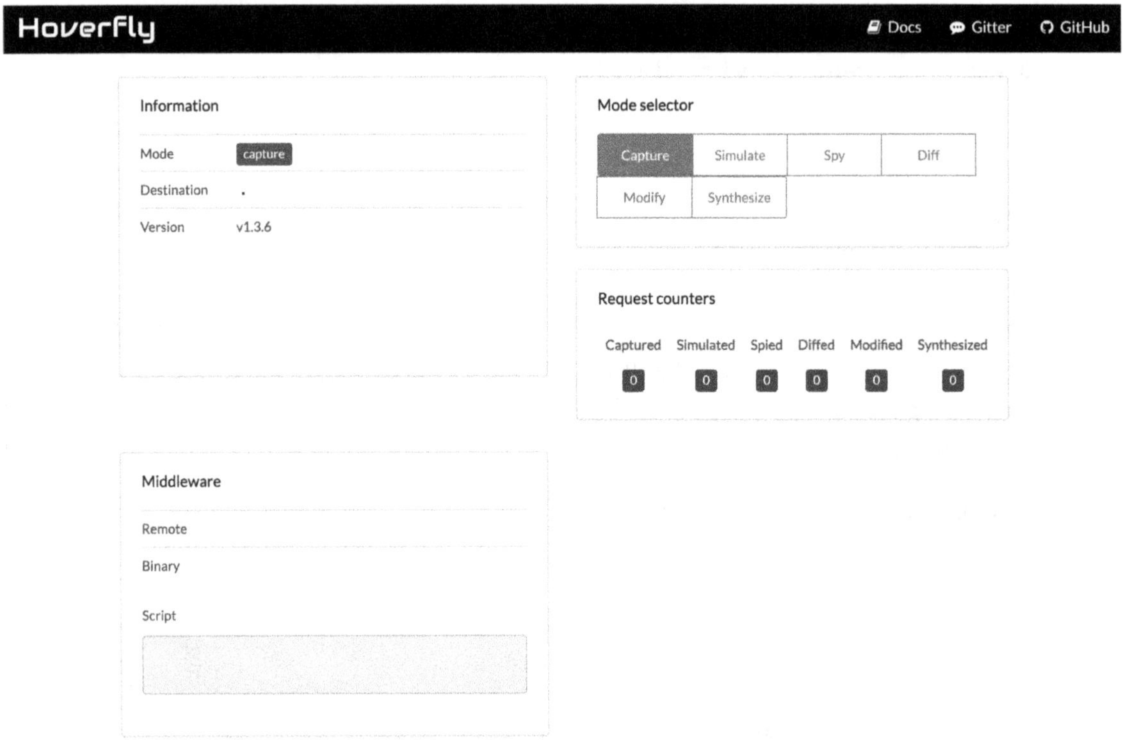

Figure 7.1 – The Hoverfly landing page

Indeed, the requests have been captured. Now we should export the data that Hoverfly has captured.

We can export all captured data into a JSON file:

```
curl http://localhost:8888/api/v2/simulation
{"data":{"pairs":[{"request":{"path":[{"matcher":"exact",
```

```
"value":"/location/0c2e2081-075d-443a-ac20-

...

"schemaVersion":"v5.1","hoverflyVersion":"v1.3.6",
"timeExported":"2022-05-22T13:35:46Z"}}
```

The simulation retrieved earlier will fetch the captured data from all the services that are subject to intercepting traffic. We will go one step further and extract the captured traffic per service.

Extracting the location service simulation

Note that `task-manager` uses `location-service`. In order to be able to run `task-manager` for testing purposes, we should simulate `location-service` using Hoverfly.

A simulation file is already placed inside the location-service directory. In order to extract a simulation from your previous request, you can follow the next step.

Extract the simulation for the `location-service` directory:

```
cd location-service
curl --location --request GET 'http://localhost:8888/api/
v2/simulation?urlPattern=location-service:8080' > location-
simulation.json
```

The `location-simulation.json` file will contain a simulation scenario that can be used by running Hoverfly in simulation mode.

Extracting the Pushgateway simulation

`event-service` issues requests to the Pushgateway in order to expose metrics. We will export the captured from the Pushgateway.

A simulation file is already placed in the `monitoring` directory. In order to extract a simulation from your previous request, you can extract the simulation for the `push-gateway` service:

```
cd monitoring
curl --location --request GET 'http://localhost:8888/api/
v2/simulation?urlPattern=push-gateway:9091' > push-gateway-
simulation.json
```

The `push-gateway-simulation.json` file will contain a simulation scenario that can be used by running Hoverfly in simulation mode.

Adapting the simulation

While in simulation mode, Hoverfly has certain rules and matchers in terms of the components of an http request. For example, in order to simulate a request for a dynamic endpoint with path variables, Hoverfly should be configured to respond with a payload provided the target endpoint matches a regex expression of an existing endpoint in the Hoverfly simulation.

In our case, the body of the REST calls will be dynamic. Therefore, we will make an adaption to the previously exported simulations and accept the payload found in the body of a POST request using "body":[{"matcher":"glob","value":"*"}]}.

Creating mock applications using Hoverfly

We can now create mock applications using the simulations we exported previously.

First, we will focus on creating a task-manager deployment using the Hoverfly simulation for location-service.

Mock location service

Now that we have the Hoverfly simulation, we are able to simulate location-service without the need to run the actual service. Our Compose deployment will be focused only on the task-manager service.

The compose file that will utilize Hoverfly for simulation will be the following:

```
services:
  location-service:
    image: spectolabs/hoverfly:v1.3.6
    ports:
      - 8888:8888
    networks:
      - location-network
      - redis-network
    volumes:
      - ./location-service/location-simulation.json:/etc/
hoverfly/location-simulation.json
    entrypoint: ["hoverfly","-webserver","-listen-on-
host","0.0.0.0", "-import", "/etc/hoverfly/location-simulation.
json", "-pp","8080"]
```

Let's run and see the results:

```
docker compose -f base-compose.yaml -f task-manager/docker-
compose.yaml -f location-service/mock-location-service.yaml up
```

We are able to interact with the task-manager service without the need to spin up the location service.

Mock Pushgateway

The next service to try to run standalone would be the event service. The component it was depending on was push-gateway. Since we have the simulation from the previous step, let's create a Docker file that would not have that dependency:

```
services:
  push-gateway:
    image: spectolabs/hoverfly:v1.3.6
    ports:
      - 8888:8888
    networks:
      - monitoring-network
      - redis-network
    volumes:
      - ./monitoring/push-gateway-simulation.json:/etc/
hoverfly/push-gateway-simulation.json
      entrypoint: ["hoverfly","-webserver","-listen-on-
host","0.0.0.0", "-import", "/etc/hoverfly/push-gateway-
simulation.json", "-pp","8080"]
```

Now we will run the event service as a standalone without depending on using the Compose files of the other services:

```
docker compose -f base-compose.yaml -f event-service/docker-
compose.yaml -f monitoring/mock-push-gateway.yaml  up
```

We are able to interact with the task-manager service without the need to spin up the location service. Also, we can run the event service without the need to run the push-gateway component. We run the applications by only using the components needed and no other services. By doing so, we are able to be flexible with our development.

Creating different environments

Previously, we managed to resolve the dependencies between our services and offer the ability to run only what we needed, which served our scenario back then.

If we examine the compose commands that we ran, we can identify that different files have been used in each case.

Compose gives us the flexibility to combine the different compose files and assemble different environments.

Running with capturing enabled

As we discovered earlier, we can have an environment for capturing the traffic exchanged between applications using Hoverfly:

```
docker compose -f base-compose.yaml -f monitoring/docker-
compose.yaml -f event-service/capture-traffic-docker-compose.
yaml -f location-service/docker-compose.yaml -f task-manager/
capture-traffic-docker-compose.yaml -f hoverfly/proxy.yaml up
```

This environment could be used when users want to create new simulations for testing.

Running with monitoring disabled

Also, we can have a lean environment without monitoring enabled:

```
docker compose -f base-compose.yaml -f monitoring/mock-push-
gateway.yaml -f event-service/docker-compose.yaml -f location-
service/docker-compose.yaml -f task-manager/docker-compose.yaml
up
```

This environment could help in running the Compose application with fewer resources.

Running applications individually

During development, focusing on one component at a time is crucial. We are now able to do so by running the services in isolation and using mocks wherever applicable:

- `task-manager`:

  ```
  docker compose -f base-compose.yaml -f location-service/
  mock-location-service.yaml -f task-manager/docker-
  compose.yaml up
  ```

- `location-service:`

  ```
  docker compose -f base-compose.yaml -f location-service/
  docker-compose.yaml up
  ```

- `event-service:`

  ```
  docker compose -f base-compose.yaml -f monitoring/mock-
  push-gateway.yaml -f event-service/docker-compose.yaml up
  ```

By having the original application split into different Compose files, it was possible for us to try different combinations of those files and have a different application outcome. By combining the different compose files that we used, we ended up with different environments serving certain purposes. For example, we can have an environment without monitoring, an environment focused on capturing traffic, or a mock environment for testing. Since we are now aware of the combinations that interest us, instead of combining the compose files manually through the command line, we can proceed with extracting a unified configuration for each case.

Combining multiple Compose files into one

We have created various environments by combining compose files. This will assist the development process; however, it will make it more complex. Compose gives us the option to combine the various compose files used for the various use cases into one.

Using config

Note that `config` is a Docker compose command that will merge the files specified.

For example, we can try it when we want to run the location service as standalone:

```
docker compose -f base-compose.yaml -f location-service/docker-
compose.yaml config
```

The result will be the unified JSON:

```
name: chapter7
services:
  location-service:
    build:
      context: /path/to/git/A-Developer-s-Essential-Guide-to-
Docker-Compose/Chapter7/location-service
      dockerfile: Dockerfile
    depends_on:
```

```
      redis:
        condition: service_started
    environment:
      REDIS_HOST: redis:6379
    healthcheck:
      test:
      - CMD
      - curl
      - -f
      - http://localhost:8080/ping
      timeout: 5s
      interval: 10s
      retries: 5
      start_period: 5s
    image: location-service:0.1
    networks:
      location-network: null
      monitoring-network: null
      redis-network: null
  redis:
    image: redis
    networks:
      redis-network: null
networks:
  location-network:
    name: chapter7_location-network
  monitoring-network:
    name: chapter7_monitoring-network
  redis-network:
    name: chapter7_redis-network
```

Here, we managed to generate a merged compose installation using `config`. This way, we have a more managed way to assemble the compose files we use in different scenarios.

Summary

We managed to transform a monolithic compose application into a flexible modular one with multiple Compose files. Also, based on the dependencies among Compose files, we proceeded with creating mock services in order to be able to run each service for development purposes in a lean way. Onward, we combined various compose files and created different environments for our application. Then, we proceeded to merge the various assembled files into one based on the scenario that needed tackling.

In the next chapter, we will see how we can simulate production environments using compose.

8

Simulating Production Locally

In the previous chapter, we managed to modularize our microservice-based application into different Compose files. Also, we went ahead with creating different environments for those applications. We have an environment with mock services, an environment that captures traffic between services, and an environment with monitoring enabled.

By being able to use mock services, generate different environments, and monitor our applications, we are able to be more productive and efficient in everyday development. In this chapter, we shall focus on simulating production locally using Compose.

A development team can be productive from the start if it has fewer dependencies and a development environment ready for testing.

Our target scenario will be an AWS environment. We shall simulate AWS services locally and also make a representation of a Lambda-based AWS environment through a Docker Compose application.

The target environment will be a simple application receiving a JSON payload. The application shall store the information in DynamoDB and then send the updates to a **Simple Queue Service** (**SQS**) queue. Another Lambda application will read the SQS messages and store them in **Simple Storage Service** (**S3**) for archival purposes.

In a real AWS environment, all the components involved, including SQS, **Simple Notification Service** (**SNS**), S3, and DynamoDB, are well integrated, thus making the operation of the application streamlined. However, having this environment available for local testing will require some workarounds to make up a well-integrated AWS environment. The components of our application will be a REST-based Lambda application storing the request in DynamoDB, an application simulating the publishing of SQS messages in a Lambda function, and the SQS-based Lambda application storing SQS events in S3.

The topics we shall cover in this chapter are the following:

- Segregating private and public workloads
- Setting up DynamoDB locally
- Setting up SQS locally

- Setting up S3 locally
- Setting up a REST-based Lambda function
- Setting up an SQS-based Lambda function
- Connecting the Lambda functions

Technical requirements

The code for this book is hosted on the GitHub repository at `https://github.com/PacktPublishing/A-Developer-s-Essential-Guide-to-Docker-Compose`. If there is an update to the code, it will be updated on the GitHub repository.

Segregating private and public workloads

Since the actions taking place in AWS are internal, we should separate the workloads into private and public.

The REST-based Lambda application receiving the JSON payload needs to be on a public network, since it will interact with the end user. The SQS-based Lambda application, reading the SQS events and storing them in S3, needs to be private. The application simulating the SQS events to the SQS-based Lambda application will also be private.

The mock AWS components, such as DynamoDB, SQS, and S3, should use the private network.

We shall define the networks with the following Compose configuration:

```
networks:
  aws-internal:
  aws-public:
```

By having the private networks defined, we can now proceed with adding the mock AWS components to the Compose application.

Setting up DynamoDB locally

A commonly used Database on AWS is DynamoDB. DynamoDB is a serverless key/value **NoSQL** database. For local testing, AWS provides us with a local version of DynamoDB.

We shall use the Docker images provided by AWS and add them to the Compose configuration. For convenience, we shall expose the port locally.

As mentioned before, the DynamoDB service will use the private network defined previously:

```
services:
  dynamodb:
    image: amazon/dynamodb-local
    ports:
      - 8000:8000
    networks:
      - aws-internal
```

Since DynamoDB locally is up and running, let's create a table on it.

Creating DynamoDB tables

Unlike Redis, in DynamoDB we need to create a table beforehand. We shall add a container to the Compose application, which creates the table in DynamoDB.

We did something similar to this in *Chapter 2, Running the First Application Using Compose*.

The container will use the AWS CLI image (`https://hub.docker.com/r/amazon/aws-cli`), override the command in order to use the DynamoDB util included, and create the table.

The initialization container will depend on the DynamoDB service, since DynamoDB needs to be available. The rest of the application will depend on the initialization service, since the table needs to exist before using it.

The script that will create the table will be the following:

```
#!/bin/sh
aws dynamodb create-table \
    --table-name newsletter \
    --attribute-definitions \
        AttributeName=email,AttributeType=S \
    --key-schema \
        AttributeName=email,KeyType=HASH \
    --provisioned-throughput \
        ReadCapacityUnits=5,WriteCapacityUnits=5 \
    --table-class STANDARD --endpoint-url  http://dynamodb:8000
http://host.docker.internal:8000
```

Then, we shall add the initialization container to the Compose application:

```
services:
  dynamodb-initializer:
    image: amazon/aws-cli
    env_file:
      - ./mock_crentials.env
    entrypoint: "/create_table.sh"
    depends_on:
      - dynamodb
    volumes:
      - ./create_table.sh:/create_table.sh
    networks:
      - aws-internal
```

As you can see, we added some mock credentials in order to use the AWS CLI and also override the endpoint for DynamoDB. We can now test DynamoDB locally.

Interacting with the Local DynamoDB

We can test the local DynamoDB we set up previously by running a small snippet.

First, let's start DynamoDB:

```
$ docker compose -f base-compose.yaml up
```

We shall use Go for running the example; therefore, we can use an existing go project, or create an example project with the initialization commands we used in the previous chapters.

Once done, we need to include the following dependencies (the following commands need to be executed from the dynamodb-snippet directory):

```
$ go get github.com/aws/aws-sdk-go/aws
$ go get github.com/aws/aws-sdk-go-v2/service/dynamodb
```

We can now use the following small snippet that puts an entry in the DynamoDB table:

```
sess, _ := session.NewSession(&aws.Config{
    Region:      aws.String("us-west-2"),
    Credentials: credentials.
NewStaticCredentials("fakeMyKeyId", "fakeSecretAccessKey", ""),
})
```

```
svc := dynamodb.New(sess, aws.NewConfig().WithEndpoint("http://
localhost:8000").WithRegion("eu-west-2"))
item := Subscribe{
    Email: "john@doe.com",
    Topic: "what I subscribed",
}
av, _ := dynamodbattribute.MarshalMap(item)
input := &dynamodb.PutItemInput{
    Item:      av,
    TableName: aws.String("Newsletter"),
}
svc.PutItem(input)
```

We have been successful in simulating using DynamoDB locally. We also managed to create a table using a container. We also ran a code example that will persist items in the DynamoDB table we created. Our Compose application has a DynamoDB running, making it possible for our services to interact with it. The next step would be to add a mock SQS component to our Compose application.

Setting up SQS locally

SQS will be used in order to notify us when a DynamoDB entry has been created. The REST-based Lambda application will send a message in SQS.

elasticmq is a very popular SQS emulator tool (https://github.com/softwaremill/elasticmq), which covers most of the features provided by SQS.

In order to push data to SQS, a queue should be created. elasticmq provides us with the option to create a queue on initialization.

The configuration will be the following:

```
//sqs.conf
include classpath("application.conf")

queues {
  subscription-event{}
}
```

Let's now add the `elasticmq` configuration to our Compose file:

```
services:
  sqs:
    image: softwaremill/elasticmq
    ports:
      - 9324:9324
      - 9325:9325
    networks:
      - aws-internal
    volumes:
        - ./sqs.conf:/opt/elasticmq.conf
...
```

As we did with DynamoDB, for convenience reasons, we shall expose the port locally. Also, `elasticmq` provides us with an administrator interface on port `9325` (`http://localhost:9325/`).

Let's interact with the local SQS broker using a Go snippet.

The following module needs to be included (the following commands need to be executed from the `sqs-snippet` directory):

```
$ go get github.com/aws/aws-sdk-go/aws
$ go get github.com/aws/aws-sdk-go/service/sqs
```

Our code snippet will print the available queues in the service:

```
session, _ := session.NewSession(&aws.Config{
    Region:      aws.String("us-west-2"),
    Credentials: credentials.NewStaticCredentials("fakeMyKeyId",
"fakeSecretAccessKey", ""),
})
svc := sqs.New(session, aws.NewConfig().WithEndpoint("http://
localhost:9324").WithRegion(os.Getenv(AWS_REGION_ENV)))

result, _ := svc.ListQueues(nil)

for i, url := range result.QueueUrls {
    fmt.Printf("%d: %s\n", i, *url)
}
```

We have successfully run an SQS simulator locally. We also created an SQS queue using the emulator's integrated functionalities in creating queues. We also implemented a code example, which was successful in publishing data to the SQS queue by using the emulator endpoint. The services hosted on Compose should be able to interact with SQS and publish messages. In the next section, we shall set up a mock S3 server on Compose to facilitate blob storage in our application.

Setting up S3 locally

S3 is a highly available object storage service provided by AWS. As with most AWS services, it provides a REST API to interact with as well as an SDK.

In order to simulate S3 locally, we shall use S3mock (`https://github.com/adobe/S3Mock`), a highly rated project on GitHub.

A Docker image is available for it, which also provides the configuration option to create a bucket from the start.

We shall add it to our Compose file and attach it to the internal network:

```
services:
...
  s3:
    image: adobe/s3mock
    ports:
      - 9090:9090
    networks:
      - aws-internal
    environment:
      - initialBuckets=subscription-bucket
```

We will add a code snippet for it; thus, the following package needs to be included (the following commands need to be executed from the s3-snippet directory):

```
$ go get github.com/aws/aws-sdk-go/aws
$ go get github.com/aws/aws-sdk-go/service/s3
```

Our code snippet will list the available buckets:

```
sess := session.Must(session.NewSessionWithOptions(session.
Options{
    SharedConfigState: session.SharedConfigEnable,
}))
```

```
s3 := s3.New(sess, aws.NewConfig().WithEndpoint("http://
localhost:9090").WithRegion("us-west-2"))
buckets, _ := s3.ListBuckets(nil)
for i, bucket := range buckets.Buckets {
    fmt.Printf("%d: %s\n", i, *bucket.Name)
}
```

We managed to run an S3 emulator locally. We configured the emulator and initialized it by using a bucket. Next, we ran a code example that will point to the local S3 bucket and list the existing buckets created. In the next section, we shall set up a REST-based Lambda function.

Setting up a REST-based Lambda function

AWS provides us with Lambda. AWS Lambda is a serverless computing offering that can be integrated and invoked in various ways. One way it can be utilized is by using it as a backend for REST APIs.

The REST-based Lambda function that we shall implement will receive a JSON payload and store it in DynamoDB.

This can be easily simulated locally, since AWS provides `docker-lambda`.

By using `docker-lambda`, we can create a container image that can simulate our AWS Lambda function. AWS provides images for this purpose that also include a runtime interface client that facilitates the interaction between our function code and Lambda (`https://github.com/lambci/docker-lambda`).

Furthermore, this makes it feasible to simulate calls to the Lambda function locally.

Let's start with the function's code base.

Initially, we shall persist the request in DynamoDB:

```
type Subscribe struct {
    Email string `json:"email"`
    Topic string `json:"topic"`
}


func HandleRequest(ctx context.Context, subscribe Subscribe)
(string, error) {
    dynamoDb, _:= dynamoDBSession()
    marshalled, _ := dynamodbattribute.MarshalMap(subscribe)
    input := &dynamodb.PutItemInput{
            Item:        marshalled,
```

```
            TableName: aws.String(TableName),
    }


    dynamoDb.PutItem(input)
    sendToSQS(subscribe)


    return fmt.Sprintf("You have been subscribed to the %s
newsletter", subscribe.Topic), nil
}
```

Then, we shall send a message to SQS:

```
func sendToSQS(subscribe Subscribe) {
    if !isSimulated() {
        return
    }


    if session, err := sqsSession(); err == nil {
        if bytes, err := jsonutil.BuildJSON(subscribe); err
== nil {
            smsInput := &sqs.SendMessageInput{
                MessageBody: aws.String(string(bytes)),
                QueueUrl:    aws.String(os.Getenv(SQS_
TOPIC_ENV)),
            }


            if _, err := session.SendMessage(smsInput);
err != nil {
                fmt.Println(err)
            }
[...]
func sqsSession() (*sqs.SQS, error) {
    session, _ := session.NewSession()


    return sqs.New(session, aws.NewConfig().WithEndpoint(os.
Getenv(SQS_ENDPOINT_ENV)).WithRegion(os.Getenv(AWS_REGION_
ENV))), nil
}
```

The full source example can be found on GitHub (https://github.com/PacktPublishing/A-Developer-s-Essential-Guide-to-Docker-Compose/blob/main/Chapter8/newsletter-lambda/newsletter.go).

Let's now create the Dockerfile for the application:

```
FROM amazon/aws-lambda-go:latest as build
RUN yum install -y golang
RUN go env -w GOPROXY=direct
COPY go.mod ./
COPY go.sum ./
RUN go mod download
COPY *.go ./
RUN go build -o /main
FROM amazon/aws-lambda-go:latest
COPY --from=build /main /var/task/main
CMD [ "main" ]
```

Everything is set up to add a Compose file for the application:

```
services:
  newsletter-lambda:
    build:
      context: ./newsletter-lambda/
    image: newsletter_lambda
    ports:
     - 8080:8080
    environment:
     - SIMULATED=true
     - DYNAMODB_ENDPOINT=http://dynamodb:8000
     - SQS_ENDPOINT=http://sqs:9324
     - SQS_TOPIC=/000000000000/subscription-event
    depends_on:
       - dynamodb-initializer
       - sqs
    env_file:
       - ./mock_crentials.env
    networks:
      aws-internal:
      aws-public:
```

As we can see, we reference the mock AWS services we created previously. Also, we build the Docker image through the Compose file. This is a public service and the entry point for our application; therefore, we expose the port locally.

Let's run it using Compose:

```
docker compose -f docker-compose.yaml -f newsletter-lambda/
docker-compose.yaml build
docker compose -f docker-compose.yaml -f newsletter-lambda/
docker-compose.yaml up
```

As we can see, we combined the Compose files as we did in *Chapter 7, Combining Compose Files*.

Our service is up and running, and we can test it by issuing a request with `curl`:

```
curl -XPOST "http://localhost:8080/2015-03-31/
functions/function/invocations" -d '{"email":"john@doe.
com","topic":"Books"}'
"You have been subscribed to the Books newsletter"
```

To sum up, we managed to create an AWS Lambda function on Compose and facilitated its interactions with the mock DynamoDB and SQS services. We managed to simulate an AWS serverless-based application through Compose without interacting with the AWS console. In the next section, we will go one step further and introduce an SQS-based AWS Lambda function to our Compose application.

Setting up an SQS-based Lambda function

Previously, we managed to run locally a REST-based AWS Lambda function. Our next component will also be a Lambda function but message-based; more specifically, it will listen to the SQS events we emitted previously.

The Lambda application, by receiving the SQS events, will then persist them in S3. The same components we used previously will also be used for this application.

Let's see the function handler:

```
func HandleRequest(ctx context.Context, sqsEvent events.
SQSEvent) error {
    session := s3Session()
    for _, message := range sqsEvent.Records {
        var subscribe Subscribe
        json.Unmarshal([]byte(message.Body), &subscribe)

        key := fmt.Sprintf("%s.%d", hash(subscribe.Email),
```

```
time.Now().UnixNano()/int64(time.Millisecond))

        marshalled, _ := json.Marshal(subscribe)

        session.PutObject(&s3.PutObjectInput{
            Bucket: aws.String(os.Getenv(SUBSCRIPTION_BUCKET_
ENV)),
                Key:    aws.String(key),
                Body:   bytes.NewReader(marshalled),
            })
        }
    return nil
}
```

The function handler will receive `SQSEvent` containing SQS messages. Each message will be unmarshalled and stored in S3 using a hash- and time-based generated key.

AWS streamlines the SQS message handling. If the function invocation is successful, the message shall be removed from SQS. If not, the message will stay in the queue.

In order to build the image, a Dockerfile is needed:

```
FROM amazon/aws-lambda-go:latest as build
RUN yum install -y golang
RUN go env -w GOPROXY=direct
COPY go.mod ./
COPY go.sum ./
RUN go mod download
COPY *.go ./
RUN go build -o /main
FROM amazon/aws-lambda-go:latest
COPY --from=build /main /var/task/main
CMD [ "main" ]
```

Due to this being a Lambda-based application, the Dockerfile is identical to the one we implemented for the REST-based Lambda application.

Next, we shall create the Compose file:

```
services:
  s3store-lambda:
```

```
build:
  context: ./s3store-lambda/
image: s3store-lambda
environment:
  - SIMULATED=true
  - S3_ENDPOINT=http://s3:9090
  - SUBSCRIPTION_BUCKET=subscription-bucket
  - AWS_REGION=eu-west-2
links:
-   "s3:subscription-bucket.s3"
depends_on:
  - s3
env_file:
  - ./mock_crentials.env
networks:
  aws-internal:
```

The application is accessed only internally; thus, it resides in the internal network. Also, we reference the mock AWS service we defined previously; however, there is a crucial detail in the `links` section.

Docker Compose links

Due to the way S3 works when accessing a bucket, instead of accessing through the root of the S3 endpoints, the bucket name is appended at the endpoints.

If the bucket name is `my-bucket`, the URL in order to interact with this bucket will be `https://my-bucket.s3.your-region.amazonaws.com/`.

This is in conflict with our deployment, since we have `s3` as our endpoint, and our code base will try to access `subscription-bucket.s3`.

To tackle this, we shall utilize the `links` functionality that Compose provides us with.

By using `links`, we can define `subscription-bucket.s3` as an extra alias for the `s3` service; therefore, we shall be able to reach it via our service.

So far, we have successfully created an SQS-based Lambda function as well as run it locally. We managed to use S3 and an alias workaround for the bucket-based endpoint. In the next section, we shall combine the two applications through an intermediate local-only application that simulates the AWS environment for SQS-based Lambda functions.

Connecting the Lambda functions

So far, we have set up the mock AWS components for S3 and SQS, and we created two Lambda functions, one for REST-based communication and one for SQS-based communication. In an AWS environment, both functions would be seamlessly integrated, since by publishing a message to SQS, AWS handles the dispatching of that message to the Lambda function that should process it.

This seamless integration is what we miss in the current state of our Compose application. In order to facilitate this functionality, we shall create a service that pulls images from SQS and pushes them to the SQS-based function.

The code base is very streamlined:

```
session, _ := sqsSession()
queueUrl := aws.String(os.Getenv(SQS_TOPIC_ENV))
msgResult, _ := session.ReceiveMessage(&sqs.
ReceiveMessageInput{
    QueueUrl: queueUrl,
})
if msgResult != nil && len(msgResult.Messages) > 0 {
    sqsEvent := map[string][]*sqs.Message{
    "Records": msgResult.Messages,
}

marshalled, _ := json.Marshal(sqsEvent)
http.Post(os.Getenv(S3STORE_LAMBDA_ENDPOINT_ENV), "application/
json", bytes.NewBuffer(marshalled))
    for i := 0; i < len(msgResult.Messages); i++ {
        session.DeleteMessage(&sqs.DeleteMessageInput{
            QueueUrl:       queueUrl,
            ReceiptHandle:
            msgResult.Messages[i].ReceiptHandle,
        })
    }
}
```

The messages will be pulled from an SQS service, formatted in the format that the Lambda function expects to receive. Once the messages have been dispatched to the Lambda function, they shall be deleted from the queue.

This service will work only locally; thus, the image creation will be much simpler.

Let's build the Dockerfile for the image:

```
# syntax=docker/dockerfile:1
FROM golang:1.17-alpine
WORKDIR /app
COPY go.mod ./
COPY go.sum ./
RUN go mod download
COPY *.go ./
RUN go build -o /main
CMD [ "/main" ]
```

Next, we shall create the Compose configuration:

```
services:
  sqs-to-lambda:
    build:
      context: ./sqs-to-lambda/
    image: sqs-to-lambda
    environment:
      - SQS_ENDPOINT=http://sqs:9324
      - SQS_TOPIC=/000000000000/subscription-event
      - S3STORE_LAMBDA_ENDPOINT=http://s3store-
lambda:8080/2015-03-31/functions/function/invocations
    depends_on:
      - sqs
      - s3store-lambda
    env_file:
      - ./mock_crentials.env
    networks:
      aws-internal:
networks:
  aws-internal:
```

The service is internal and will use only SQS. Since it will execute requests to s3store-lambda, it is dependent on it.

> **Note**
>
> If you have any active Compose sessions, ensure that they are stopped before moving on and executing the commands that will follow next.

Let's run the entire application and see how the services interact together:

```
docker compose -f docker-compose.yaml -f newsletter-lambda/
docker-compose.yaml -f s3store-lambda/docker-compose.yaml -f
sqs-to-lambda/docker-compose.yaml build
docker compose -f docker-compose.yaml -f newsletter-lambda/
docker-compose.yaml -f s3store-lambda/docker-compose.yaml -f
sqs-to-lambda/docker-compose.yaml up
```

Let's invoke the REST-based Lambda function the same way we did previously:

```
curl -XPOST "http://localhost:8080/2015-03-31/
functions/function/invocations" -d '{"email":"john@doe.
com","topic":"Books"}'
"You have been subscribed to the Books newsletter"
```

We should be able to see logs on all services by now:

```
...
chapter8-newsletter-lambda-1      | START RequestId: f2dcc750-
35a1-40d8-9c54-f7c2edc3bcfe Version: $LATEST
chapter8-newsletter-lambda-1      | END RequestId: f2dcc750-
35a1-40d8-9c54-f7c2edc3bcfe

...


chapter8-sqs-to-lambda-1          | 2022/07/24 21:31:03
Dispatching 1 received messages
...
chapter8-s3store-lambda-1         | START RequestId: 7caff9ab-
ddb4-46c7-b75f-0f726eaf2ae8 Version: $LATEST
chapter8-s3store-lambda-1         | END RequestId: 7caff9ab-
ddb4-46c7-b75f-0f726eaf2ae8
```

Through this internal service, we managed to simulate a functionality that AWS provides out of the box. The limitations that we had initially were resolved by a Compose-driven solution. Being the entry point for our application, the REST-based service stores data on DynamoDB and sends messages to SQS. The SQS messages are then transmitted to the SQS-based Lambda function using this internal service.

Summary

In this chapter, we managed to spin a cloud-based infrastructure locally on our workstation in a seamless way. We configured the equivalent mock components for the AWS services DynamoDB, SQS and S3. Through our Compose configuration, we managed to configure them and also tackle some limitations that happen during local development. This gave us the option to develop our code base upon those services without the need to interact with an actual production environment.

Next, we proceeded to implement services suitable for the AWS Lambda environment. We successfully run those Lambda functions through our Compose application while making them eligible for deployment to a cloud environment. Last but not least, we simulated some functionality that AWS provides by introducing a local private application. Through the course of this chapter, there was no need to interact with the AWS console and a real production environment, and the focus remained on the development of the code base.

In the next chapter, we shall take advantage of this chapter's code base and use Compose for our **Continuous Integration and Continuous Deployment (CI/CD)**.

9

Creating Advanced CI/CD Tasks

In the previous chapter, we managed to simulate an AWS environment locally through a Compose application. We mocked AWS services such as DynamoDB, S3, and SQS. Also, we simulated the invocation of Lambda functions through Docker containers and came up with a workaround to simulate traffic toward SQS-based Lambda services by introducing an extra service in the Compose installation.

This enabled us to be focused on developing our application without the need to interact with the AWS console, provision any AWS infrastructure, and deal with the needs of a cloud-hosted environment. From the beginning, we were focused on developing the application locally and simulating the components needed.

Since we have been productive so far in developing the application, the next logical step is to introduce some CI/CD to our current application. Throughout the development life cycle, we want our application to build, test, and deploy automatically.

Our Lambda-based application is a good example of how we can benefit from Compose and simulate a complex application on the chosen CI/CD solution. The Lambda application requires more than one component to operate in order to test. Compose can assist in spinning up this environment in the CI/CD solution of our choice.

In this chapter, the focus will be on enabling Docker Compose in a CI/CD solution. When it comes to CI/CD, there are various vendors and software packages out there. Therefore, we will examine more than one CI/CD solution.

In this chapter, we will cover the following topics:

- Introduction to CI/CD
- Using Docker Compose with GitHub Actions
- Using Docker Compose with Bitbucket pipelines
- Using Docker Compose with Travis

Technical requirements

The code for this book is hosted on the GitHub repository at `https://github.com/PacktPublishing/A-Developer-s-Essential-Guide-to-Docker-Compose`. In the case of an update to the code, it will be updated to the GitHub repository.

Introduction to CI/CD

CI/CD stands for **continuous integration** and **continuous delivery**. It is a combination of practices that facilitate continuous integration, continuous delivery, and continuous deployment. Part of its scope is the automation of building, testing, and deploying applications.

For example, let's take our Lambda application. It is a complex environment consisting of two Lambda-based applications and three different AWS services.

For our use case, we assume that we have a team that follows trunk-based development, a practice that facilitates CI. Our team will contribute small commits to the trunk-main branch every time. This can be done with short-lived feature branches. Pull requests will be raised in order to merge changes from those branches to the trunk-master branch. A pull request should be reviewed by the development team, in parallel, a CI/CD automated process that builds and tests the newly introduced code should take place and be part of the merge check. Once the merge checks have been passed, our branch is ready to be merged and a deployment of the component should happen.

Regardless of the component that we change, whether it is a REST-based Lambda function or an SQS-based Lambda function, we need to make sure that the changes on that function will not break our code base and the applications that interact with that component.

Once a merge takes place, the component that we have merged should be built and then pushed to a live environment. Deploying code to a live environment can vary based on where the workloads are getting deployed. For example, a Lambda function deployment requires a new Docker image and an invocation of the AWS API to point to the Docker image we have built. This would also require some extra configuration based on the environment that an AWS Lambda function can have. If, in the future, we switch to Kubernetes for deploying the application, a Helm chart can be deployed manually, or a GitOps solution such as Argo CD can be adopted. A GitOps solution will poll for changes that took place on the trunk branch, pick the latest build artifact, and deploy it to the live environment without any user intervention.

We want to be feasible for our application to use Compose in CI/CD tasks. The deployment canary can either be the Go binary deployed to the Lambda function or a Docker image. Also, in the case of changing the code base in the future, another environment might require a different deployment. Therefore, we will ignore the deployment and focus on enabling the execution of Compose commands for the required CI/CD jobs.

For every commit that happens on our main branch, we will spin up the Compose application on the CI/CD job that will be triggered. Our goal is to make the entire Compose application run on a CI/CD build and test the application before we proceed to deployment.

We have an overview of CI/CD and what we want to achieve in terms of our Lambda-based application. Since the source code is hosted on GitHub, we will proceed with implementing CI/CD jobs for our application using GitHub Actions.

Using Docker Compose with GitHub Actions

If your code base is hosted on GitHub, it is highly like that you are aware of GitHub Actions. GitHub Actions is the CI/CD platform provided by GitHub. By using GitHub Actions, we can add workflows that build and test our code base. This can be adapted for each branch and pull request or be used to add custom workflows and deploy our code base through GitHub.

Creating your first GitHub Action

In order to add a GitHub workflow, you need to place YAML files along with workflow instructions inside the `.github/workflows` directory. Multiple files can be added, and they should be executed by GitHub independently.

For now, we will focus our app to execute on the main branch.

This is the base of our workflow:

```
name: subscription-service

on:
  push:
    branches:
    - main
jobs:
  build:
    timeout-minutes: 10
    runs-on: ubuntu-latest
    steps:
    - name: Checkout
      uses: actions/checkout@v2
```

The name of our workflow is `subscription-service`, and the workflow will be executed once there is a push on the main branch.

The virtual environment to use is `ubuntu-latest`, which is provided by GitHub Actions. The benefit of using this environment is that Compose comes preloaded with it.

Then, we add a step to check out the repository. Since each job will take place on a new instance of a virtual environment, we should be concerned about the dependencies and the artifacts that get produced in each step. For this case, a cache mechanism is provided by GitHub so that we can speed up the time it takes to load these dependencies.

Caching built images

Building images can be time-consuming. This is something we want to avoid in CI/CD since we want our jobs to be fast and as smooth as possible.

Long-running jobs can have a negative impact on the automation process:

- The job can time out, thus making it impossible to be invoked.
- The development process gets slower.
- CI/CD becomes painful for developers.

For this reason, we will use the caching capabilities provided by GitHub Actions:

```
...
    - name: Cache Local Images
      id: local-images
      uses: actions/cache@v3
      with:
        path: /var/lib/docker/
        key: local-docker-directory
...
```

By adding this step, we instruct `actions` to cache the `/var/lib/docker` directory. This is the directory where images are stored. By doing so, the steps that will interact with this directory will have the content-generated cache; therefore, the next jobs will be able to pick up the artefacts that have already been downloaded from the previous steps.

Building application images

We are ready to add our next step, which will be to build the application images. As we saw in *Chapter 4, Executing Docker Compose Commands*, we can use Compose to build the application images.

The action step should be the following:

```
...
    - name: Build Images
      working-directory: ./Chapter8
      run: |
          docker compose -f docker-compose.yaml -f newsletter-
      lambda/docker-compose.yaml -f s3store-lambda/docker-compose.
      yaml -f sqs-to-lambda/docker-compose.yaml build
...
```

Since we focused on the application we developed in *Chapter 8, Simulating Production Locally*, we should switch the working directory to the corresponding directory. We will use the `working-directory` section pointing to the `Chapter8` directory.

In the `run` section, we specify the `build` command that we used previously. This can be any `bash` command available through the virtual environment chosen.

The outcome of this action will be the application's Docker images to be built.

Since we have built the images, we are ready to proceed with making a proof-of-concept test for our application.

Testing your Compose application

It's time to add our next Compose step, which will be a proof-of-concept test. We will spin up the entire Compose application and then run a `curl` command, as we did earlier, and check the results.

The step we are going to add is the following:

```
...
    - name: Test application
      working-directory: ./Chapter8
      run: |
          docker compose -f docker-compose.yaml -f newsletter-
      lambda/docker-compose.yaml -f s3store-lambda/docker-compose.
      yaml -f sqs-to-lambda/docker-compose.yaml up -d
          sleep 20
          curl -XPOST "http://localhost:8080/2015-03-31/
      functions/function/invocations" -d '{"email":"john@doe.
      com","topic":"Books"}'
          sleep 20
```

```
        docker compose logs --tail="all"
    ...
```

We set up `Chapter8` as the working directory. Then, we spin up the Compose application in daemon mode. Daemon mode gives us the ability to continue to use the Terminal session while the application is running.

By sleeping for 20 seconds, we can make sure every service is running. Then, a `curl` command will invoke the Lambda function serving as an entry point.

Since the actions are async, we will wait for some seconds.

Provided our command is successful, we can then check for the logs using the Compose logs command with the `-tail` option.

Overall, we managed to run the entire Compose application in a pipeline. Also, we did manage to run a test. This will make our automation efforts more efficient since we can use Compose to simulate a prod-like infrastructure in CI/CD, interact with it, and apply automated checks during the development process. Onward, we will implement the same job on Bitbucket pipelines.

Using Docker Compose with Bitbucket pipelines

Bitbucket pipelines are a CI/CD solution for the repositories hosted on Bitbucket. By having a repository hosted on Bitbucket instead of using an external solution for CI/CD purposes, Bitbucket pipelines can be very useful since they are readily available and seamlessly integrated. As with the previous case in GitHub Actions, we will follow the same process.

Creating your first Bitbucket pipeline

In order to enable Bitbucket pipelines, you need to create a `bitbucket-pipelines.yml` file in the root directory of your project. Then, you can enable the pipelines in your repository through the settings:

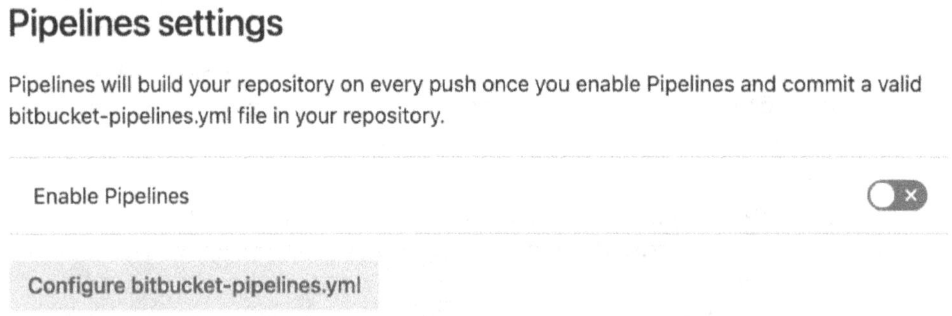

Pipelines settings

Pipelines will build your repository on every push once you enable Pipelines and commit a valid bitbucket-pipelines.yml file in your repository.

Enable Pipelines   ⬤✕

Configure bitbucket-pipelines.yml

Figure 9.1 – Bitbucket pipelines enabled

Once the pipelines are enabled, Bitbucket will proceed with executing the instructions specified in the bitbucket-pipelines.yml file.

The bitbucket-pipelines.yml base will be the following:

```
image: atlassian/default-image:3

options:
  docker: true

definitions:
  caches:
    compose: ~/.docker/cli-plugins

pipelines:
  default:
    - step:
        name: "Install Compose"
        caches:
          - compose
        script:
          - mkdir -p ~/.docker/cli-plugins/
          - curl -SL https://github.com/docker/compose/
releases/download/v2.2.3/docker-compose-linux-x86_64 -o
~/.docker/cli-plugins/docker-compose
          - chmod +x ~/.docker/cli-plugins/docker-compose
          - docker compose version
```

In the image section, we specify the Docker image that we should use throughout the execution of the CI/CD tasks. The atlassian/default-image:3 image is based on a Linux distribution that is more specific to Ubuntu 20.04 LTS. By default, this image does not have Compose support; therefore, we will have to install Compose in a Linux environment. We follow the same step that we followed in *Chapter 1, Introduction to Docker Compose*, in order to install Compose on a Linux distribution.

Also, as we can see, in the pipeline, we enable the Docker services. This is crucial to enable access to the Docker daemon and, thus, be able to interact through Docker commands. By enabling Docker services through these options, we enable their Docker capabilities for all of the pipeline steps.

Caching Compose and Docker images

Our first step was to add support for Compose to the Docker image used. Since in every step, a new Docker container is being created, we need to execute the same commands for each pipeline step. To prevent this, we will cache the Docker plugin directory. Once the next step takes place, the directory and its contents will be present through the cache, making it feasible to use Compose.

So, what we did in the `caches` section is to add a custom cache for Bitbucket pipelines, pointing to a directory.

Take note that although this is a workaround for the current pipelines, there is a more efficient workaround for building an image based on `atlassian/default-image:3`, which will have the Compose installation instructions executed.

Instead of creating custom caches, we can use existing implementations. Bitbucket comes with various caches predefined, and one of them is for Docker. By having Docker caching enabled on a pipeline step, we can make sure that the images that are already downloaded and the images built will be cached and ready to be used for the next steps.

The caches to be used in each step are specified using the `caches` section:

```
...
    - step:
        name: "Hello world"
        caches:
          - docker
        script:
          - docker run --rm hello-world
...
```

In this case, the image for `hello-world` will be downloaded and cached.

By having caching enabled for Compose and Docker, we can speed up the process of the pipeline steps without the need to retrieve the dependencies needed in each step. Therefore, we will proceed to the next step, which is building the images.

Building application images

We have been able to execute Compose commands in a Bitbucket pipeline environment. Now, we can proceed with interacting with our Compose application using Compose commands.

We will add a step in the pipeline that builds the images:

```
...
    - step:
        name: "Chapter 8 Build Images"
        caches:
          - docker
          - compose
        script:
          - cd Chapter8
          - docker compose -f docker-compose.yaml -f
newsletter-lambda/docker-compose.yaml -f s3store-lambda/docker-
compose.yaml -f sqs-to-lambda/docker-compose.yaml build
...
```

There is not much difference compared to the previous GitHub Actions example; we switched to the directory manually and issued the same build command.

As we can see, we have the docker and compose caches enabled. Therefore, there is no need to install Compose again. Also, the images that we built will be available for the next pipeline step. By having the images built, we can proceed with executing a test for our application.

Testing your Compose application

The images have been built, so now we can test the Compose application as we did earlier.

The step for testing the application is as follows:

```
...
    - step:
        name: "Chatper 8 Test Application"
        caches:
          - docker
          - compose
        script:
          - cd Chapter8
          - docker compose -f docker-compose.yaml -f
newsletter-lambda/docker-compose.yaml -f s3store-lambda/docker-
compose.yaml -f sqs-to-lambda/docker-compose.yaml up -d
          - sleep 20
```

```
            - curl -XPOST "http://localhost:8080/2015-03-31/
functions/function/invocations" -d '{"email":"john@doe.
com","topic":"Books"}'
            - sleep 20
            - docker compose logs --tail="all"
    ...
```

We will have the same outcome that we had with GitHub Actions. Docker Compose is set up in Bitbucket pipelines. This enables us to proceed with more advanced CI/CD tasks in Compose that utilize Compose. Now we can proceed with implementing this logic on another popular CI/CD provider, Travis CI.

Using Docker Compose with Travis

Travis is a YAML-based CI/CD solution. It does provide source code hosting, but it is very well integrated with GitHub. Travis used to provide free CI/CD for open source projects, so it is very common to work on an open source project that uses Travis. The same steps we followed in the previous CI/CD vendors will also be applied to Travis.

Creating your first Travis job

Travis is YAML-based just like the previous CI/CD tools we examined. Once we enabled Travis integration with a GitHub project, a file named .travis.yml containing the job instructions needs to be present at the root location of the project.

Our .travis.yml base should be the following:

```
services:
  - docker

cache:
  directories:
    - $HOME/.docker/cli-plugins

jobs:
  include:
    - stage: "Install Compose"
      script:
        - mkdir -p /home/travis/.docker/cli-plugins/
```

```
        - curl -SL https://github.com/docker/compose/releases/
  download/v2.2.3/docker-compose-linux-x86_64 -o ~/.docker/cli-
  plugins/docker-compose
        - chmod +x ~/.docker/cli-plugins/docker-compose
        - docker compose version
```

The job specification seems familiar. As we did with the Bitbucket pipelines, here, we also installed the Compose binary for the Linux distribution. Therefore, caching this step is as essential as before.

Caching Compose

Travis has caching capabilities. As we can see, we cached the directory where Compose is installed. By specifying the cache, it will take effect for all the jobs included in our configuration.

From the online documentation, Travis CI discourages the caching of Docker images. Therefore, we won't proceed in this direction. However, if caching is needed, certain images can be cached by saving and loading the images through a cached directory.

Building application images

By being able to execute Compose commands on Travis, we can now proceed with building the images.

Here is the job that will build the images:

```
...
      - stage: "Build Images"
        script:
        - cd Chapter8
        - docker compose -f docker-compose.yaml -f newsletter-
  lambda/docker-compose.yaml -f s3store-lambda/docker-compose.
  yaml -f sqs-to-lambda/docker-compose.yaml build
...
```

Since the job is successful the next step is to test the application.

Testing your Compose application

Our testing stage will not have any significant difference from the ones we implemented earlier apart from the syntax used.

The test section will be the following:

```
...
      - stage: "Test application"
```

```
      script:
        - cd Chapter8
        - docker compose -f docker-compose.yaml -f newsletter-
lambda/docker-compose.yaml -f s3store-lambda/docker-compose.
yaml -f sqs-to-lambda/docker-compose.yaml up -d
        - sleep 20
        - curl -XPOST "http://localhost:8080/2015-03-31/
functions/function/invocations" -d '{"email":"john@doe.
com","topic":"Books"}'
        - sleep 20
        - docker compose logs --tail="all"
  . . .
```

As expected, the images have been built and tested successfully. So, we are able to use Compose on multiple CI/CD providers and enhance our automation pipeline.

Summary

We did it! We managed to run CI/CD tasks that utilized the functionalities of Compose. This way, complex production environments could be simulated through CI/CD allowing us to have a fine-grained way of integrating the code base and a sufficient amount of merge checks.

In the next chapter, we will check how we can use Compose to deploy to a remote host.

Part 3:
Deployment with Docker Compose

This part will focus on how to benefit from Docker Compose on production deployments. Infrastructure on public clouds such as AWS and Azure will be provisioned and Compose deployments will be deployed upon that infrastructure. Lastly, we will also see how we can migrate our Compose workloads to the Kubernetes orchestration engine.

The following chapters will be covered under this section:

10

Deploying Docker Compose Using Remote Hosts

In the previous chapter, we created CI/CD tasks by using Docker Compose. We also created various environments that we can use and utilize regarding the scenario presented.

In this chapter, we will focus on deploying our Docker applications to a remote host. While developing an application, there are various reasons why you may not want to deploy your application to another host: the application can be resource-intensive, you may want to share the progress with a colleague or the host, the application is getting deployed to, may have access to resources through the network that your workstation doesn't. A remote host could be a solution to those issues since it allows us to deploy a Docker application to another workstation and thus make it available externally.

In this chapter, we will cover the following topics:

- Docker remote hosts
- Creating a remote Docker host
- Docker Contexts
- Deploying Compose to remote hosts
- Executing remote host deployments through your IDE

Technical requirements

The code for this book can be found in the following GitHub repository: `https://github.com/PacktPublishing/A-Developer-s-Essential-Guide-to-Docker-Compose`. If any updates are made to the code, they will be reflected in the GitHub repository.

Docker remote hosts

Imagine an application running locally on a machine and you want it to be accessible by another individual through a Linux machine located on the cloud for general usage. If this application is based on Docker Compose, the application can be deployed manually to the VM through shell commands. However, there is a more streamlined way to deploy this application to the target VM. Provided a server has Docker installed, it is eligible to become a Docker host. Docker gives you the ability to use the Docker capabilities of another machine, provided it has Docker installed and access to that machine has been set up.

An example of a remote host is the Docker installation on Windows and macOS. They both require a Linux VM to run Docker. The Linux VM is the remote host that the Docker CLI uses to interact with.

Now, let's learn how to create a remote host.

Creating a remote Docker host

To create a Docker host, we need a Linux machine. This can even be a spare laptop or a spare VM that runs a Linux distribution. The provisioning commands are the same commands we followed in *Chapter 1, Introduction to Docker Compose*. Since a spare Linux workstation may not be available, we shall create a Docker host using AWS EC2.

Creating a Docker host on AWS EC2

In this section, we shall spin up a machine on AWS using EC2. This instance will become our remote host. These steps apply to any available Linux-based server, so the EC2 part can be skipped if you have a Linux workstation available.

By navigating to the IAM section of the AWS console, we should retrieve a key and a secret. This key and secret need to belong to a user that can provision an EC2 machine:

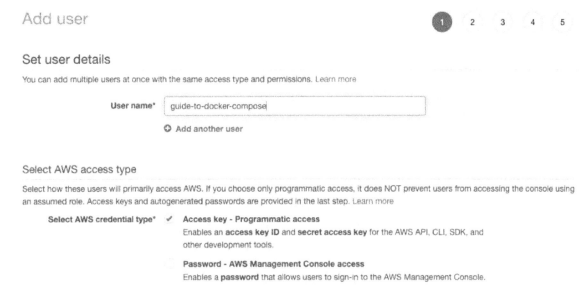

Figure 10.1 – AWS user

Once we've retrieved the credentials, we can proceed to the VPC section to find the default VPC for the region selected:

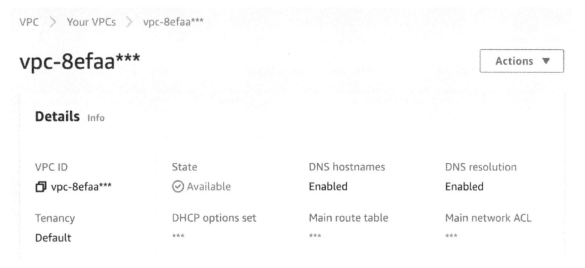

Figure 10.2 – VPC network

Copy that VPC ID since we'll need it later.

To streamline the provisioning of EC2, we shall use Terraform.

Installing Terraform

Terraform is a modern **Infrastructure as Code** (**IaC**) solution. Infrastructure and resources on the cloud can be defined by using a declarative configuration language.

To install Terraform on your system, you can follow the instructions in the official documentation (`https://learn.hashicorp.com/tutorials/terraform/install-cli`).

Once the Terraform binary is present in the command line, we can check its version, as follows:

```
$ terraform version
Terraform v1.2.3
on darwin_arm64
```

Terraform provisions your infrastructure and keeps track of the changes in the Terraform state. The Terraform state can be a local file, a file hosted on AWS S3 with the equivalent blob solutions of other cloud providers, or it can be customized provided the user creates a plugin for it. For example, it is possible to store the state in a database such as RavenDB, provided you develop a plugin for it. In our case, we will just use the local filesystem to store the state.

When running Terraform, it will pick up the cloud provider used and download the binaries needed. For example, if we provision code for AWS using Terraform, Terraform will download the AWS plugins without us having to do any extra installation work.

Setting up an EC2 machine with SSH enabled

The goal is to set up an EC2 instance that we can use to log in using SSH. This instance needs to have Docker and Docker Compose installed.

We would like this machine to only be accessible from our workstation's IP. Thus, we should provide our IP when provisioning the infrastructure. Also, the EC2 machine will be in a virtual private network. In our case, we want to use the default VPC. To do so, we shall use the ID of the default VPC – the one that we copied previously.

We should specify the IP and the VPC ID, as variables so that we can use them when we provision the infrastructure:

```
variable "myvpc" {
}
variable "myip" {
}
```

Now, let's generate those SSH keys that we will use for the EC2 machine.

Once the keys have been generated, the private key should be added to the OpenSSH authentication agent.

The command to generate the keys and add them to the OpenSSH authentication agent:

```
// Chapter10/generate-key.sh
$ ssh-keygen -t rsa -b 2048 -f $(pwd)/ssh.key -N ""
$ ssh-add ssh.key
```

We need to execute this step before provisioning the EC2 instance to provision the EC2 machine using an existing key. Also, by adding this key to the SSH authentication agent, we streamline the process of connecting to the server we will be provisioning.

The next step is defining the infrastructure. Since we will SSH to that machine, we need a security group that will allow ingress to the instance from our workstation.

The ingress rule for this is as follows:

```
resource "aws_security_group" "remote_docker_host_security_
group" {
   ...
   ingress {
      description       = "SSH from workstation"
      from_port         = 22
      to_port           = 22
      protocol          = "tcp"
      cidr_blocks       = ["${var.myip}/32"]
   }
   ...
}
```

As you can see, we use the `ip` variable we specified previously.

We also need to enable egress. If we are going to have a remote host on that machine, we need to be able to interact with external Docker registries:

```
resource "aws_security_group" "remote_docker_host_security_
group" {
   ...
   egress = [
      {
```

```
        cidr_blocks      = [ "0.0.0.0/0", ]
        description      = ""
        from_port        = 0
        ipv6_cidr_blocks = []
        prefix_list_ids  = []
        protocol         = "-1"
        security_groups  = []
        self             = false
        to_port          = 0
      }
    ]

    ...

}
```

Having generated the keys, AWS gives us the option to upload the public key as a resource. This can make the procedure involve more bootstrapping and will add the SSH key to multiple machines, which may act as Docker hosts.

The SSH key resource is as follows:

```
resource "aws_key_pair" "docker_remote_host_key" {
  key_name   = "docker-remote-host-key"
  public_key = file("${path.module}/ssh.key.pub")
}
```

An EC2 machine will be created and will use the key we created previously.

Finally, we must create the EC2 instance:

```
resource "aws_instance" "remote_docker_host" {
    ami = "ami-078a289ddf4b09ae0"
    instance_type = "t2.micro"


    key_name = aws_key_pair.docker_remote_host_key.key_name


    vpc_security_group_ids = [
        aws_security_group.remote_docker_host_security_group.id
    ]
}
```

This EC2 instance will be provisioned. It will allow traffic from our workstation, and we will be able to have access from the outside.

However, this will require us to install Docker on the machine that's running. Instead, we can use the user-data functionality of an EC2 machine. user-data is the script that runs once an EC2 machine has been provisioned.

By utilizing this functionality, we can set up Docker Compose on the EC2 machine:

```
resource "aws_instance" "remote_docker_host" {
...
  user_data = <<-EOF
    #!/bin/bash
    yum install docker -y
    usermod -aG docker ec2-user
    systemctl start docker
    su ec2-user
    mkdir -p /home/ec2-user/.docker/cli-plugins
    curl -SL https://github.com/docker/compose/releases/
download/v2.2.3/docker-compose-linux-x86_64 -o /home/ec2-user/.
docker/cli-plugins/docker-compose
    chmod +x /home/ec2-user/.docker/cli-plugins/docker-compose
  EOF
...
}
```

The preceding commands should seem familiar to you; we ran them in *Chapter 1, Introduction to Docker Compose*. Since we picked up a Red Hat-based VM image, we used yum.

Since we'll be connecting to that machine, let's also print the EC2 machine's IP address:

```
output "instance_ip" {
  description = "Remote host ip"
  value       = aws_instance.remote_docker_host.public_ip
}
```

Now, we have everything we need to provision the infrastructure.

To execute the necessary Terraform commands, we can pass the credentials needed for AWS through the `AWS_ACCESS_KEY_ID` and `AWS_SECRET_ACCESS_KEY` environment variables. We must also specify the region we will operate in via `AWS_REGION`. These variables can be exported or passed directly to the Terraform command. You can also use them indirectly by using the credentials files and configuration that was generated when configuring `aws-cli`.

First, let's initialize Terraform:

```
AWS_ACCESS_KEY_ID=key-id AWS_SECRET_ACCESS_KEY=access-key AWS_
REGION="eu-west-2" terraform init
```

This will provide our state in a file.

Now, run the following command:

```
AWS_ACCESS_KEY_ID=*** AWS_SECRET_ACCESS_KEY=*** AWS_REGION="eu-
west-2" terraform apply -var myip=51.241.***.182 -var
myvpc=vpc-a8d1b***
```

By doing this, the infrastructure will be provisioned. Based on the output, we shall get the IP needed to SSH:

```
instance_ip = "18.133.27.148"
```

Now, we can SSH using the key we created previously and check that Docker Compose exists:

```
$ ssh ec2-user@18.130.80.179
[ec2-user@ip-172-31-37-105 ~]$ docker compose version
Docker Compose version v2.2.3
```

As expected, the user data script runs successfully. We were also able to `ssh` to the instance since the keys have been set up.

Using the remote Docker host

Now that the remote host is available, let's see how we can make it execute a Docker command using the host.

Let's try to run Redis using the host:

```
DOCKER_HOST="ssh://ec2-user@18.130.80.179" docker run -it --rm
redis
```

If we log into the EC2 instance and execute `docker ps`, we shall see that `redis` is running on that machine:

```
$ ssh ec2-user@18.130.80.179 docker ps
CONTAINER ID    IMAGE       COMMAND                 CREATED
STATUS          PORTS       NAMES
e44e3bd3a41d    redis       "docker-entrypoint.s…"  10 seconds
ago     Up 9 seconds    6379/tcp    nifty_aryabhata
```

By creating a Docker remote host, we can create Docker containers on that host by using our local workstation. This opens new possibilities since more than one host can be used. Managing hosts can be demanding. Therefore, in the next section, we'll learn how to achieve this using Docker Contexts.

Docker Contexts

Using the host on each command that we use is redundant and error-prone. For example, a deployment may fail due to it reaching a different host because we omitted to specify the host when running the command and we executed a different command on our local host.

For this case, Docker Contexts can be of help.

By creating contexts, we can switch our Docker configuration to multiple contexts and pick the right context per case.

So, let's create a context for our EC2 host:

```
$ docker context create ec2-remote --docker host=ssh://
ec2-user@18.130.80.179
```

Although we have created the context, we are still in the default context. Let's switch to the recently created context:

```
$ docker context use ec2-remote
```

Run the following command:

```
$ docker run -it --rm redis
```

Check the results on the server:

```
ssh ec2-user@18.130.80.179 docker ps
CONTAINER ID    IMAGE       COMMAND                 CREATED
STATUS          PORTS       NAMES
```

```
1b5b0459bf48     redis        "docker-entrypoint.s…"     15 seconds
ago     Up 13 seconds    6379/tcp     peaceful_feynman
```

However, there is no need to run this command on the server. Thanks to using the context of EC2-remote, we can use the docker ps command locally and the results will be the same. The context will take effect until we switch contexts again:

```
$ docker ps
CONTAINER ID       IMAGE        COMMAND                       CREATED
STATUS             PORTS        NAMES
1b5b0459bf48       redis        "docker-entrypoint.s…"       3 minutes ago
Up 3 minutes       6379/tcp     peaceful_feynman
```

By, having Docker Contexts enabled, we can use Docker Compose on a remote host.

Deploying Compose to remote hosts

Let's move the Redis example we run previously into a Compose file:

```
services:
  redis:
    image: redis
```

The following command will have to be adjusted with regards to the DOCKER_HOST variable, since a different IP will be allocated to the EC2 instance. The outcome should be the same on another host:

```
$ DOCKER_HOST="ssh://ec2-user@18.130.80.179" docker compose up
[+] Running 2/2
 Network chapter10_default     Created
0.1s
 Container chapter10-redis-1   Created
0.1s
Attaching to chapter10-redis-1
chapter10-redis-1  | 1:C 22 Jun 2022 22:50:52.725 #
oO0oo0OoO000o Redis is starting oO0oo0OoO000o
```

By checking the host, we should see that a Redis instance is running.

Since we just used Docker Contexts, we don't need to specify the host. So, let's try one more time without the DOCKER_HOST environment:

```
$ docker context use ec2-remote
$ docker compose up
[+] Running 2/2
...
```

Here, we ran our application using a remote host on an EC2 machine. Now, we need to clean up the infrastructure we provisioned and keep the costs minimized. We can destroy the infrastructure manually through the AWS console, but since we provisioned the infrastructure using Terraform, we can use the `destroy` command it provides.

Let's clean up our infrastructure:

```
AWS_ACCESS_KEY_ID=*** AWS_SECRET_ACCESS_KEY=*** AWS_REGION="eu-
west-2" terraform destroy -var myip=51.241.***.182 -var
myvpc=vpc-a8d1b***
```

The use cases we've covered so far can provide a solid developer experience. The necessary code is being developed, environments are being provisioned and deployed through Compose, and we can deploy the application to a remote host, making it accessible to other users. The next step will be to enhance our development efforts by deploying to a Docker host straight from our IDE.

Executing remote host deployments through your IDE

When developing an application, an **integrated development environment** (**IDE**) has a crucial role in making us more productive. By using Compose, we can deploy and simulate environments, which is why it has become part of our day-to-day development. In this section, we shall combine the usage of an IDE and Compose.

In this section, we will use IntelliJ IDEA Ultimate Edition (https://www.jetbrains.com/idea/download/#section=mac) as our IDE. Ultimate Edition comes with the option of a free trial.

Let's configure the Docker host. First, go through the **Preferences** section, then the **Build**, **Execution**, **Deployment**, and **Docker** sections. Now, a new Docker configuration can be added:

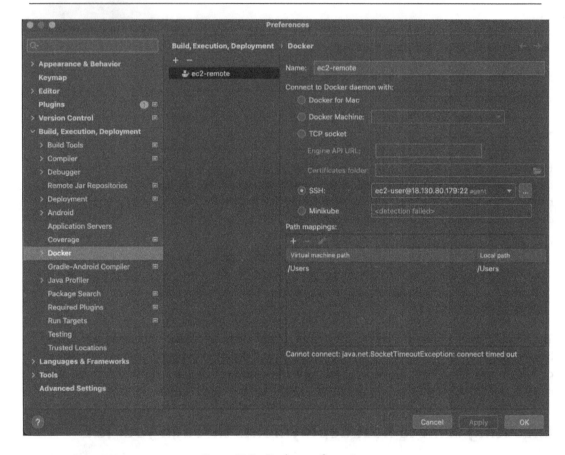

Figure 10.3 – Docker configuration

Then, provided we have `docker-compose.yaml`, we can run it locally:

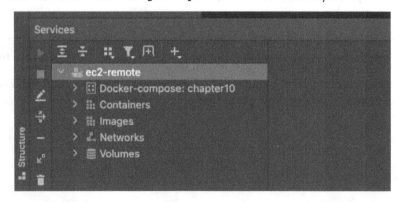

Figure 10.4 – Running Compose

As a result, when we run the Compose file from our IDE, it will use the remote host:

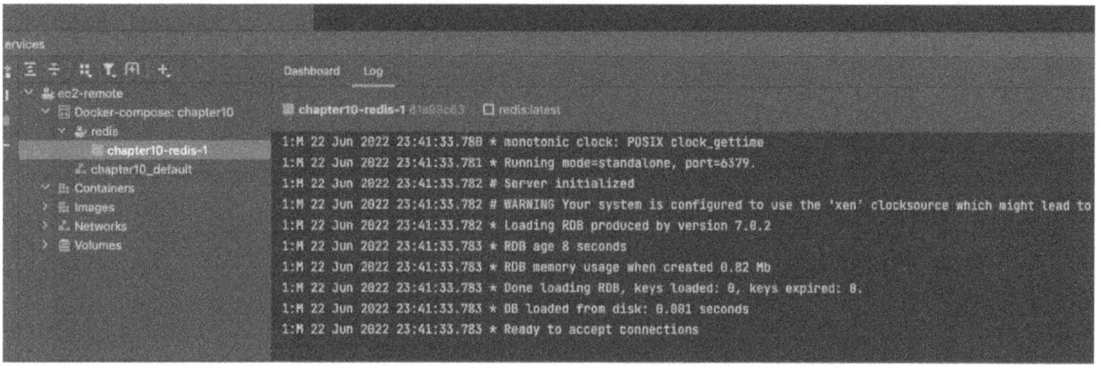

Figure 10.5 – Logs

So, apart from developing our application locally, we managed to deploy it to a remote host and make it feasible for other individuals to check out our progress.

Summary

In this chapter, we deployed our Compose applications to a remote server. This helped us utilize remote server resources and share our application through a remote server. By doing so, we deployed an application to a server. However, this is not a suitable way to deploy a Compose application to production.

In the next chapter, we will learn how to deploy Compose applications to the cloud using the necessary tools and make them production-ready.

11
Deploying Docker Compose to AWS

In the previous chapter, we deployed our application to a Docker host. The feature of deploying to a remote host could help in many ways, for example, we could share the application with another individual or use the remote host for development and testing purposes. Deploying to a remote host brings us closer to the context of deploying to production. However, a deployment to a remote host is not up to the standards of a production deployment. A production deployment needs our application to be highly available, secure, and accessible through a load balancer, and the logs of the application need to be easy accessible and securely stored.

This chapter is all about bringing our Docker Compose application to production. **Elastic Container Service (ECS)** is one of the container orchestration services that AWS provides. ECS is integrated with Docker Compose, therefore by using an existing Compose application we can have a cloud-native application deployed through ECS.

We will start by pushing our Docker images to the AWS **Elastic Container Registry (ECR)**. Then, we shall apply some minor adjustments to an existing Compose application to use the images, from the registry provisioned, and deploy to ECS using a Docker profile for AWS. Once we have deployed our application, we shall proceed to more advanced concepts, such as using an existing cluster in a private network, as well as scaling and secret management.

Here are the topics we will cover in this chapter:

- Introduction to AWS ECS
- Hosting your Docker images on AWS ECR
- Deploying your application to an ECS cluster
- Adapting Compose files for ECS deployment
- Running your Compose application to existing infrastructure
- Advanced Docker Compose concepts on ECS

Technical requirements

The code for this book is hosted on the GitHub repository at `https://github.com/PacktPublishing/A-Developer-s-Essential-Guide-to-Docker-Compose`. If there is an update to the code, it will be updated in the GitHub repository.

In order to provision AWS resources through the command line, it is essential to install the AWS CLI tool at `https://docs.aws.amazon.com/cli/latest/userguide/getting-started-install.html`. The AWS CLI tool can help us with administrative tasks as well as troubleshooting issues on our AWS infrastructure.

In this chapter, Terraform plans will be used. To install Terraform on your system, follow the instructions from the official documentation (`https://learn.hashicorp.com/tutorials/terraform/install-cli`).

Introduction to AWS ECS

ECS is a container orchestration engine that is provided by AWS. We can use ECS in order to deploy, manage, and scale our container applications. Since it is provided by AWS, it is integrated with the rest of the AWS platform. By deploying an application on ECS it will use an elastic load balancer to expose an application; it will use EC2 instances to run the application, and the application will reside on an AWS **Virtual Private Cloud** (**VPC**) and its subnets. The logs of the applications will also be accessible through CloudWatch.

ECS comes with the option of AWS Fargate. AWS is a serverless compute option that enables you to deploy your Docker workloads without needing to manage EC2 instances and autoscaling groups. If the application's workloads are small, require low overhead, and have non-frequent bursts of requests and usage, then Fargate is a solution that our application can benefit from. We will choose Fargate for our application, since our application is still in the prototype phase.

For applications with more demanding workloads, or workloads that have certain needs, we can back ECS with EC2 and autoscaling groups. For example, if an application is CPU intensive, it makes sense to create an autoscaling group that uses a compute-optimized EC2 type.

Working locally with containers is easy, and concerns such as image distribution might not be present in the beginning. However, to use a container that we created, it is essential to be able to distribute the image of that container. Instead of creating and provisioning a registry, AWS provides us with a container registry. In the next section, we shall push our images to an AWS-hosted container registry.

Hosting your Docker images on AWS ECR

Previously, we have been building images for Compose applications that would later be stored and retrieved from the Docker host. To use the Docker images on ECS, a Docker Image registry is essential. The registry to use on AWS is ECR. By using ECR, we push and pull the images from the registry and use them on any workstation or server that has Docker installed, provided we have configured access to that registry.

ECR is a fully managed container registry. It is a highly available solution backed by AWS **Simple Storage Service (S3)**, thus images are stored across multiple systems. Also, it has features such as scanning the images for vulnerabilities. By using a managed container registry, the maintenance overhead is reduced; for example, there is no need to provision a server or plan the storage capacity.

Another benefit of ECR is how well integrated it is with the rest of the AWS services. An ECS server or a Lambda function, provided they have the right AWS IAM permissions, can pull images seamlessly without any extra configuration.

Since ECR is going to be the Docker registry that we shall use, let's proceed and provision our registry.

Provision ECR using AWS CLI

With the AWS CLI, provisioning an ECR is a one-liner:

```
$ aws ecr create-repository --repository-name developer-guide-
to-compose-ecr
{
    "repository": {
...
    "    "repository"ri": "111111111111.dkr.ecr.eu-west-1.
amazonaws.com/ developer-guide-to-compose-e"r ",
...
    "    "imageScanningConfigurat"on": {
        "    "scanOnP"sh": false
        },
    "    "encryptionConfigurat"on": {
        "    "encryptionT"pe":  "AES"56"
        }
    }
}
```

We can get some valuable information from the output. For example, we shall use `repositoryUri` to tag our images, `imageScanningConfiguration` scans the images for vulnerabilities when enabled, and `encryptionConfiguration` is how the images we push are encrypted at rest. `repositoryUri` has a number prefix. In the preceding example, it is `111111111111`. This should be your AWS account number, therefore it will be different whenever another AWS account is being used.

Since we managed to create the registry from the command line, we can also try and create the registry using Terraform.

Be aware that you might want to delete the registry first if you want to create it again:

```
$ aws ecr delete-repository --repository-name developer-guide-
to-compose-ecr
```

Onward! We shall provision the ECR we need using Terraform.

Provision ECR using Terraform

In the previous chapter, Terraform helped us to provision the infrastructure without much effort. We can use Terraform to create the Docker registry.

This is how we can define the ECR registry:

```
resource "aws_ecr_repository" "developer_guide_to_compose_ecr"
{
  name                 = "developer-guide-to-compose-ecr"
  image_tag_mutability = "MUTABLE"

  image_scanning_configuration {
    scan_on_push = true
  }
}
```

We can now initialize the Terraform plan:

```
$ AWS_ACCESS_KEY_ID=*** AWS_SECRET_ACCESS_KEY=*** AWS_DEFAULT_
REGION=eu-west-1 terraform init
```

And then proceed to execute the Terraform plan:

```
$ AWS_ACCESS_KEY_ID=*** AWS_SECRET_ACCESS_KEY=*** AWS_DEFAULT_
REGION=eu-west-1 terraform apply
```

These commands will handle the Terraform state locally. However, for an application that is deployed publicly, and various engineers interact with it or create infrastructure for it, we want a better way of storing and sharing the state.

Storing a Terraform state file

When applying the Terraform plan commands for testing purposes, storing the state locally can be an option. However, for production use, the state should reside in the form of storage that can be accessible by other individuals. In our case, we shall store the Terraform state in an S3 bucket.

Before creating a bucket on AWS, we need to be aware of the bucket naming rules on AWS. A bucket name should be unique across all AWS accounts. This means that if an AWS account has created a bucket with a name you want to use, it will not be possible, and another name should be used.

Let's create the bucket:

```
$ AWS_ACCESS_KEY_ID=*** AWS_SECRET_ACCESS_KEY=*** AWS_DEFAULT_
REGION=eu-west-1 aws s3api create-bucket --bucket developer-
guide-to-compose-state --region eu-west-1 --create-bucket-
configuration LocationConstraint=eu-west-1
$ AWS_ACCESS_KEY_ID=*** AWS_SECRET_ACCESS_KEY=*** AWS_DEFAULT_
REGION=eu-west-1 aws s3api put-bucket-versioning --bucket
developer-guide-to-compose-state --versioning-configuration
Status=Enabled
```

Since the bucket is created, we can configure Terraform to use it to store the state:

```
//provider.tf
terraform {
  required_providers {
    aws = {
      source  = "hashicorp/aws"
      version = "~> 4.16"
    }
  }
  backend "s3" {
    bucket = "developer-guide-to-compose-state"
    region = "eu-west-1"
    key = "terraform.tfstate"
  }
```

```
    required_version = ">= 1.2.0"
}


provider "aws" {
  region  = "eu-west-1"
}
```

Since this bucket has already been created on S3, the name is reserved. When you apply the script, another bucket name should be chosen before you create your bucket.

Before running `terraform apply`, it is worth making sure that the changes that will take place are the ones we want. We can use `terraform plan`, which helps us to preview the changes we will apply to our infrastructure:

```
$ AWS_ACCESS_KEY_ID=*** AWS_SECRET_ACCESS_KEY=*** AWS_DEFAULT_
REGION=eu-west-1 terraform plan
```

Now that we are confident with the changes that will take place, we shall apply the Terraform plan:

```
$ AWS_ACCESS_KEY_ID=*** AWS_SECRET_ACCESS_KEY=*** AWS_DEFAULT_
REGION=eu-west-1 terraform apply
```

Also, you can benefit from enabling versioning on the bucket. Apart from the Git commit history, versioning can be another way to have a history of the state changes.

Since the container registry has been created, we can proceed and push the images to the registry.

Pushing images to ECR

Before interacting with a registry, we need to set up authentication.

To authenticate to ECR, we shall use the AWS CLI:

```
aws ecr get-login-password --region eu-west-1|docker login
--username AWS --password-stdin 111111111111.dkr.ecr.eu-west-1.
amazonaws.com
```

This will generate a token that will be used by the Docker client to communicate with the registry. As we can see in the first section of the command, the token is generated and in the second section after the pipe, the output of the command is used to log in to the registry we created previously.

We can now test that it works and push an image to the registry:

```
$ docker tag nginx 111111111111.dkr.ecr.eu-west-1.amazonaws.
com/developer-guide-to-compose-ecr:ngnix
```

```
$ docker push 111111111111.dkr.ecr.eu-west-1.amazonaws.com/
developer-guide-to-compose-ecr:ngnix
f2089ca22bc1: Pushed
9e13ccef5ed0: Pushed
9dfe3def52f1: Pushed
7b11943dbe46: Pushed
80730baf8465: Pushed
5978b6b69f17: Pushed
ngnix: digest:
sha256:ec2290b7c5d15abb4b3384ad66a89e9c523a4668c057898f3114fa6
1df4a5586 size: 1570
```

We tagged the Nginx image and pushed it to the registry. By using the registry, we can share our images with a container orchestration engine such as ECS.

Adapting the Compose application images

The application to use would be the Task Manager we created in *Chapter 5*, *Connecting Microservices*. We will modify the Compose file to push the images to ECR.

We should change the images on each service:

```
services:
  location-service:
...
    image: 111111111111.dkr.ecr.eu-west-1.amazonaws.com/
developer-guide-to-compose-ecr:location-service_0.1
...
  event-service:
    image: 111111111111.dkr.ecr.eu-west-1.amazonaws.com/
developer-guide-to-compose-ecr:events-service_0.1
...
  task-manager:
...
    image: 111111111111.dkr.ecr.eu-west-1.amazonaws.com/
developer-guide-to-compose-ecr:task-manager_0.1
...
```

The images we created previously are dependent on the platform our workstation is running on. ECS can support various platforms. In our case, in order to simplify the deployment, we shall pick the linux/amd64 platform.

Let's specify the platform to use before building:

```
services:
  location-service:
    platform: linux/amd64

    ...
  event-service:
    platform: linux/amd64

    ...
  task-manager:
    platform: linux/amd64

...
```

We can now build and push the images using Compose:

```
$ docker compose build --no-cache
$ docker compose push
```

We used the --no-cache option. This is essential to make sure that the images that are built are going to be built on the platform we specified. Also, it will not use any cached images from our previous builds. By using compose push, all our images will be pushed to the ECR registry.

We can now proceed and deploy the application to an ECS cluster.

Deploying your application to an ECS cluster

Deploying to ECS is seamless; however, we need to create a profile on Docker that will use AWS credentials to be able to interact with AWS and provision resources.

Therefore, we need to create a Docker Context that's specific to AWS ECS scenarios:

```
$ docker context create ecs guide-to-compose
? Create a Docker context using:  [Use arrows to move, type to
filter]
> An existing AWS profile
  AWS secret and token credentials
  AWS environment variables
? Select AWS Profile  [Use arrows to move, type to filter]
  default
> guide-to-docker-compose
Successfully created ecs context "guide-to-compose"
```

By using the credentials, the context will try to provision infrastructure or evaluate if infrastructure already exists for the Compose application. Thus, it is important for the user/role behind the AWS profile to have sufficient permissions for those actions.

Behind the scenes, Compose will generate and apply a CloudFormation template, provisioning the necessary infrastructure to facilitate running the application.

We can check the CloudFormation template, since we have export functionality in Compose, by using the `convert` command:

```
$ docker --context=guide-to-compose compose convert
AWSTemplateFormatVersion: 2010-09-09
Resources:
  CloudMap:
    Properties:
      Description: Service Map for Docker Compose project aws
      Name: aws.local
      Vpc: vpc-8efaaceb
...
```

CloudFormation is an infrastructure as code solution similar to Terraform.

An example file can be found on `https://github.com/PacktPublishing/A-Developer-s-Essential-Guide-to-Docker-Compose/tree/main/Chapter11/compose-cloudformation.yaml`.

By applying a CloudFormation template, newly defined infrastructure will be created, or existing infrastructure will be deleted or updated, depending on the changes. Essentially, based on the services and the various elements defined through the Compose application, a CloudFormation file will be generated that in turn will be applied to the existing stack created previously.

Be aware that in the previous command, we specified `context`. As we saw in *Chapter 10, Deploying Docker Compose Using Remote Hosts*, we can set `context` permanently. This is done for convenience purposes since we will not do as many administrative actions; instead, we shall focus on deployments.

We are now ready to deploy the Compose application:

```
$ docker --context=guide-to-compose compose -f compose.backup.
yaml up
⊞ LocationserviceService        CreateComplete       56.0s
⊞ EventserviceService           CreateComplete       76.1s
⊞ TaskmanagerTCP8080Listener    CreateComplete       2.0s
⊞ TaskmanagerService            CreateComplete       55.1s
```

. . .

This will create the Compose application and the infrastructure.

We can also go to CloudFormation and check the progress of the application (https://eu-west-2.console.aws.amazon.com/cloudformation/ home?rfilteringText=&viewNested=true&hideStacks=false#/ stacks?filteringStatus=active&filteringText=&viewNested= true&hideStacks=false).

We can see the progress of our application while it's being created:

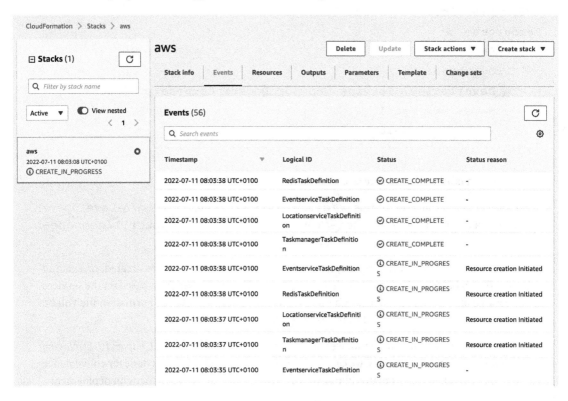

Figure 11.1 – CloudFormation progress

As we can see, the name of the stack is aws. This is because it uses the same name that a Compose deployment will use; therefore, in our case it is the name of the directory we are currently located in.

We can change this by using the project name flag:

```
$ docker --context=guide-to-compose compose -p guidetocompose
up
```

Once the application has been deployed, we can check the running containers by using `docker compose ps`:

```
$ docker --context=guide-to-compose compose ps
NAME                                            COMMAND
  SERVICE                 STATUS          PORTS
task/aws/337ff364e53e4a59b14bdaf65e4fe655       " "
  redis                 Running
task/aws/6f08bff5e5ee42a0b2d9995124debd8c       " "
location-service        Running
task/aws/bf76ab48000d49b38017e6b3b4d6b073       " "
  event-service         Running
task/aws/cea75425182c4e08ad571efcce0c82f2       " "
  task-manager          Running
aws-LoadBal-1Q44XRW9WNYF5-5fdfe7e98727ac85.elb.eu-west-1.
amazonaws.com:8080:8080->8080/tcp
```

As we can see, the Task Manager application that we exposed the port of is running and is accessible through a DNS entry.

Let's issue a test request:

```
$ curl aws-LoadBal-1Q44XRW9WNYF5-5fdfe7e98727ac85.elb.
eu-west-1.amazonaws.com:8080/ping
pong
```

Since we have achieved our goal, we will proceed with deleting the infrastructure:

```
$ docker --context=guide-to-compose compose -f compose.backup.
yaml down
```

We did manage to deploy our Compose application to AWS ECS. We did not have to provision any infrastructure since it was created by Compose using the credentials we configured through the context.

This can be a way to deploy our application; however, it forces us to create a new cluster for each application. This causes various billing concerns, and it is not the optimum way to run applications since it can lead to a maintenance overhead. In the next section, we will deploy Compose to an existing ECS cluster.

Running your Compose application to an existing cluster

Previously, we managed to run a Compose application on ECS by using an ECS Docker `context`. By deploying the application, a new infrastructure was provisioned through CloudFormation and an entire new ECS cluster was created for the application.

If we take our time and check the CloudFormation file, we can see that various AWS components have been created:

- A VPC and its subnets
- A CloudWatch log group
- Security groups
- A load balancer
- CloudMap for service discovery
- An ECS cluster
- ECS tasks

By default, CloudFormation will use the default VPC and subnets that already exist in our AWS account. The load balancer, security groups, and CloudMap, which assist with service discovery, will have to be created, as well as the ECS cluster and the ECS tasks. Those applications will be deployed to AWS Fargate.

It is obvious that these resources are provisioned and we have no control over their settings. There might be a business use case where we would like to have a private network. Also, we might want more strict egress and ingress rules.

Compose provides the option to deploy the application to an existing cluster and use resources that we have already provisioned.

In the following subsections, we will create an ECS cluster and then make adjustments to our existing application so it will use the existing infrastructure.

We will be using Terraform to provision the infrastructure.

Creating a log group

We want to keep our logs for our application in one Cloudwatch group. Therefore, we will create a log group and a log stream dedicated to our application:

```
resource "aws_cloudwatch_log_group" "task_api" {
  name              = "/ecs/task-api"
}
```

```
resource "aws_cloudwatch_log_stream" "cb_log_stream" {
  name            = "tasl-log-stream"
  log_group_name = aws_cloudwatch_log_group.task_api.name
}
```

Let's now move on to a very important aspect, which is networking.

Creating a private network

We will create a private network that will span across the specified availability zones, therefore creating a subnet per availability zone.

The Terraform plan for the VPC:

```
resource "aws_vpc" "compose_vpc" {
  cidr_block = "172.17.0.0/16"
  enable_dns_hostnames = true
  enable_dns_support = true
}
```

As we can see, we have created the VPC and specified a `cidr_block` value. Also, we enabled DNS hostnames and DNS support. This is crucial for our ECS application since communication between services requires this option to be enabled.

Let's create a subnet per availability zone:

```
resource "aws_subnet" "private_subnet" {
  count = length(var.availability_zones)
  cidr_block      = cidrsubnet(aws_vpc.compose_vpc.cidr_
block, 8, count.index)
  availability_zone = var.availability_zones[count.index]
  vpc_id          = aws_vpc.compose_vpc.id
}
resource "aws_subnet" "public_subnet" {
  count = length(var.availability_zones)
  cidr_block      = cidrsubnet(aws_vpc.compose_vpc.cidr_
block, 8, length(var.availability_zones) + count.index)
  availability_zone = var.availability_zones[count.index]
  vpc_id          = aws_vpc.compose_vpc.id
}
```

The region is specified when we run the `terraform` command. In a file containing the variables, we can have the chosen availability zones:

```
variable "availability_zones" {
  type = list(string)
  default = [ "eu-west-1a" ,"eu-west-1b" ]
}
```

Our application might need to have access to the internet. For example, we use the Redis image, which we did not deploy on the ECR registry we created. To enable access to the internet, we need an internet gateway, and we need to specify a route table that will route traffic to the internet gateway:

```
resource "aws_internet_gateway" "internet_gateway" {
  vpc_id = aws_vpc.compose_vpc.id
}
resource "aws_route" "internet_route" {
  route_table_id          = aws_vpc.compose_vpc.main_route_
table_id
  destination_cidr_block = "0.0.0.0/0"
  gateway_id             = aws_internet_gateway.internet_
gateway.id
}
```

This will work for the instances deployed to the public subnet; however, it won't work for instances that reside on the private subnet. To enable connectivity to the internet, we should use a NAT gateway.

A NAT gateway needs a public IP to operate, thus we create two for each availability zone:

```
resource "aws_eip" "nat_ips" {
  count       = length(var.availability_zones)
  vpc         = true
  depends_on = [aws_internet_gateway.internet_gateway]
}
```

A NAT gateway will be set for each public subnet:

```
resource "aws_nat_gateway" "nat_gateway" {
  count           = length(var.availability_zones)
  subnet_id       = element(aws_subnet.public_subnet.*.id, count.
index)
  allocation_id = element(aws_eip.nat_ips.*.id, count.index)
```

```
}
resource "aws_route_table" "private_route_table" {
  count   = length(var.availability_zones)
  vpc_id = aws_vpc.compose_vpc.id

  route {
    cidr_block      = "0.0.0.0/0"
    nat_gateway_id = element(aws_nat_gateway.nat_gateway.*.id,
count.index)
  }
}
```

Finally, we define a route table association for the private subnets:

```
resource "aws_route_table_association" "private_association" {
  count             = length(var.availability_zones)
  subnet_id         = element(aws_subnet.private_subnet.*.id,
count.index)
  route_table_id = element(aws_route_table.private_route_
table.*.id, count.index)
}
```

This is the most important part as it enables us to use a private network on ECS and have network connectivity.

Security groups

Security groups allow ingress and egress traffic. The full security group configuration can be found on GitHub. An important parameter is to enable connectivity between two Compose services.

To enable ingress between Compose services, use this code:

```
resource "aws_security_group_rule" "allow_services_
connectivity" {
  type                    = "ingress"
  from_port               = 0
  to_port                 = 0
  protocol                = "-1"
  source_security_group_id = aws_security_group.compose_
security_group.id
```

```
    security_group_id          = aws_security_group.compose_
  security_group.id
  }
```

This way, Compose services will be able to communicate with each other.

Configuring the ECS cluster and the load balancer

Since networking ingress and egress rules are in place, we can now configure the load balancer and the ECS cluster:

```
resource "aws_alb" "compose_alb" {
  name            = "guide-to-compose-load-balancer"
  subnets         = aws_subnet.public_subnet.*.id
  security_groups = [aws_security_group.lb.id]
}
resource "aws_ecs_cluster" "compose_ecs" {
  name = "guide-to-compose-ecs"
}
```

Updating the Compose file

Let's adapt the Compose configuration so it can use the existing cluster, VPC, and load balancer:

```
x-aws-vpc: "vpc-0144f03210f0da8e5"
x-aws-cluster: "guide-to-compose-ecs"
x-aws-loadbalancer: "guide-to-compose-load-balancer"

services:
  location-service:
    ...
    logging:
      options:
        awslogs-group: "/ecs/task-api"
  redis:
    ...
    logging:
      options:
        awslogs-group: "/ecs/task-api"
```

```
task-manager:
    ...
    logging:
      options:
        awslogs-group: "/ecs/task-api"
    links:
      - "redis:redis"
...
```

As we can see, we specified the VPC in which our workloads will be located, the ECS cluster that will orchestrate our application's containers, and the load balancer that will serve as the entry point to our application. We are now ready to deploy the application to the recently provisioned cluster.

Running your Compose application on existing infrastructure

We have provisioned the infrastructure, and we have adjusted our Compose application so that it will use the existing infrastructure. The command to provision our infrastructure will be the same as before, and the Docker profile used will be the one we used previously.

```
Run the application:$ docker --context=guide-to-compose compose
-f ./compose.aws.yaml -p guidetocompose up -d
```

Check the application containers:

```
$ docker --context=guide-to-compose compose -f ./compose.aws.
yaml -p guidetocompose ps
task/guide-to-compose-ecs/13ddf3ab03f146cc95b59005c308c5ad    ""
redis              Running
task/guide-to-compose-ecs/af645c4f87ad4a9f8d3aac5fd313bdc5    ""
location-service    Running
task/guide-to-compose-ecs/d96338ba685a401394ba9d61b80238e6    ""
task-manager           Running              guide-to-compose-load-
balancer-1956561308.eu-west-1.elb.amazonaws.com:80:80->80/http
$ curl guide-to-compose-load-balancer-1956561308.eu-west-1.elb.
amazonaws.com:80/ping
```

We did it! Our application is now running on a private network using the ECS cluster we defined using Terraform.

When you don't need the infrastructure anymore, don't forget to decommission the ECS cluster and the other resources provisioned since it will lead to extra costs.

You can do this using `terraform destroy`:

```
$ AWS_ACCESS_KEY_ID=*** AWS_SECRET_ACCESS_KEY=*** AWS_DEFAULT_
REGION=eu-west-1 terraform destroy
```

Or you can do it manually by visiting the AWS console.

Since we took more advantage of AWS components, we can also explore more advanced concepts of a Compose application deployed on ECS. Let's move on to updating and scaling our application, as well as configuring secrets.

Advanced Docker Compose concepts on ECS

We have managed to deploy our Compose application to a dedicated VPC and a dedicated ECS cluster. This gives us more control over our application and the resources we use. Building a cloud-native application comes with various benefits. We can have a rolling application update without any downtime, we can tune and scale an application based on its requirements, and we can manage and share secrets efficiently.

Updating the application

The `docker compose up` command is sufficient to update the application. Hence, the same command that is used to spin up the services will be used to update them.

Since Compose is backed by CloudFormation, the update will take place, but it can also cause downtime if a certain component of the infrastructure is removed and recreated. Rolling updates need to be configured. By using a rolling update, the container instances of a service are updated incrementally. This ensures that the application can still service requests while the containers are being updated with the latest version one at a time. We should dive deeper into scaling an application and configuring a rolling update.

Scaling the application

Since the application is deployed on an orchestration engine, we can scale up the replicas of a service. We will have two replicas of each service, except for Redis, and will configure a rolling update.

Let's put scaling instructions on all services:

```
deploy:
  mode: replicated
  replicas: 2
  update_config:
    parallelism: 1
```

```
      delay: 10s
      order: start-first
```

By increasing the replicas, the service will have two instances and ECS will handle the internal load balancing. Also, by using `update-config`, we ensure we update one service at a time.

By using `compose ps`, we can see an increase in the number of running Docker containers:

```
$ docker --context=guide-to-compose compose -f ./compose.aws.
yaml -p guidetocompose ps
NAME
COMMAND                 SERVICE                 STATUS
PORTS
task/guide-to-compose-ecs/505c7b2c962642459a8b156c4c069f73    ""
redis                   Running
task/guide-to-compose-ecs/50c74375216842dba62fea2c19772e55    ""
location-service        Running
task/guide-to-compose-ecs/8b84d785dee14441bb4a3b1808b02419    ""
event-service           Running
task/guide-to-compose-ecs/ac84fe91790244d1836238a40890c54d    ""
event-service           Running
task/guide-to-compose-ecs/b42f9c9b18704dfb888f97112b343e0b    ""
task-manager            Running                 guide-to-compose-load-
balancer-1757643662.eu-west-1.elb.amazonaws.com:80:80->80/http
task/guide-to-compose-ecs/bcce928092294fc1bcc57875954307bc    ""
task-manager            Pending                 guide-to-compose-load-
balancer-1757643662.eu-west-1.elb.amazonaws.com:80:80->80/http
task/guide-to-compose-ecs/cfd3370bfb61467e9ec15a2a1fc0763c    ""
location-service        Running
```

Instead of manually increasing the number of replicas, autoscaling can be configured:

```
deploy:
  mode: replicated
  replicas: 2
  update_config:
    parallelism: 1
    delay: 10s
    order: start-first
  x-aws-autoscaling:
    min: 2
```

```
        max:  3
        cpu:  75
```

We did it! We have a minimum of 2 replicas that will scale up to 3 if over 75% of the allocated CPU is utilized.

Using secrets

Secrets are a big part of our application. Access to a database or cloud resources needs to be configured properly. Compose gives us the option to create secrets and mount them to containers. We will mount a secret to the Task Manager:

```
    secrets:
      - secret-file
    command:
      - /bin/sh
      - -c
      - |
        ls /run/secrets/secret-file
        /task_manager
  networks:
    location-network:
    redis-network:
  secrets:
    secret-file:
      file: ./secret.file.txt
```

We also changed command of the container, before we run the task-manager application, we ls the secret-file, thus we justify that the secret is indeed mounted and resides on the container. By default, secrets are mounted on /run/secrets/secret-name.

If we check the console, the file path is printed:

/run/secrets/secret-file [GIN-debug] [WARNING] Creating an Engine instance with the Logger and Recovery middleware already attached.

In a production environment, secrets need to be handled very carefully. A local file mounted as a secret might need to be encrypted. KMS can be used for encrypting secrets and decrypting them once they've been deployed during initialization. For example, **sops** (https://github.com/mozilla/sops) is a tool that can be used for securely storing secrets on a repository using KMS.

Summary

In this chapter, we managed to deploy our Compose application to an AWS environment. We created a container registry on AWS and pushed the Docker images to the registry. We then deployed our application to ECS and a new set of infrastructure was provisioned for our application. Then we created a private network and an ECS cluster. Our Compose application has the security benefits of a private network and our infrastructure is reusable for other Compose applications. This was achieved by adapting our Compose file and specifying the infrastructure to be used. We moved on to more advanced deployment concepts, such as autoscaling and storing secrets. By adapting the Compose application, we took advantage of ECS' autoscaling capabilities and the rolling update functionality, and we shared our secrets among multiple applications.

In the next chapter, we shall deploy our Compose application to another popular cloud provider, Microsoft Azure.

12

Deploying Docker Compose to Azure

In the previous chapter, we deployed our application to AWS. We deployed the application using autoscaling and health checks, and we even managed to access the application through a DNS domain. In this chapter, we shall focus on another popular cloud provider, Azure. Azure provides us with **Azure Container Instances** (**ACI**), a seamless way to run Compose-based applications without managing any infrastructure. Deploying to ACI is simple; first, we shall push the application images to an Azure container registry, and then, with a few adjustments, we shall deploy our application to Azure ACI.

Deploying to ACI comes with less infrastructure maintenance overhead as well as with the simplicity of the Compose syntax.

We will cover the following topics in this chapter:

- An introduction to ACI
- Pushing to an Azure container registry
- Adapting Compose files for Azure container groups
- Deploying your Compose application to Azure container groups

Technical requirements

The code for this book is hosted on the GitHub repository at `https://github.com/PacktPublishing/A-Developer-s-Essential-Guide-to-Docker-Compose`. If there is an update to the code, it will be updated on the GitHub repository.

In this chapter, Terraform plans will be used. To install Terraform on your system, you can follow the instructions from the official documentation (`https://learn.hashicorp.com/tutorials/terraform/install-cli`).

Also, `az` , the Azure command-line interface, needs to be installed. The instructions can be found in the official documentation (`https://docs.microsoft.com/en-us/cli/azure/`).

An introduction to ACI

ACI is a serverless solution provided by Microsoft Azure. ACI makes it possible to deploy containers without managing any servers and infrastructure.

A benefit of ACI is that you can run multiple containers without the need to deal with the complexity of a Docker orchestration engine.

However, ACI has some limitations. Based on the region, there are some limits on the maximum number of CPUs (four CPUs for every region) as well as on the available memory. More information on the limitations can be found in the docs (`https://docs.microsoft.com/en-us/azure/container-instances/container-instances-region-availability`). Another limitation is the fact that scaling on the containers is not available (`https://docs.microsoft.com/en-us/azure/container-instances/container-instances-faq#how-do-i-scale-a-container-group-`).

It is important to take the preceding into consideration, since they play a crucial role in understanding whether our application is suitable for ACI. In our case, our application can benefit from ACI since it is easy to deploy and has simple requirements.

To deploy to ACI, we need to have our images accessible in a way so that ACI will retrieve them. A registry must be in place, so we will use the Docker registry that Azure provides. Next, we will create an Azure container registry.

Pushing to an Azure container registry

An Azure container registry will play a significant role in our deployment. It will be a universally available registry well integrated with the Azure infrastructure that we will provision.

A basic step to creating resources on Azure is to manage them under a resource group. Think of an Azure resource group as a container that includes all the resources needed for our application. It is a way to logically separate infrastructure. For example, deleting a resource group will lead to the deletion of the resources provisioned under this resource group.

Our Azure container registry as well as our ACI will reside in a resource group that we will provision. This can be done in many ways – for example, through the command line, the Azure portal, as well as using Terraform.

Adding a resource group via Terraform should look like this:

```
resource "azurerm_resource_group" "guide_to_docker_compose_
resource_group" {
```

```
    name      = "guidetodockercompose"
    location = "eastus"
}
```

We picked `eastus` on purpose, since it has more features and fewer limitations than other regions when it comes to ACI.

Now that we have the resource group in place, we can create a registry:

```
resource "azurerm_container_registry" "guide_to_docker_compose_
registry" {
    name                  = "developerguidetocomposeacr"
    resource_group_name = azurerm_resource_group.guide_to_docker_
compose_resource_group.name
    location              = azurerm_resource_group.guide_to_docker_
compose_resource_group.location
    sku                   = "Basic"
    admin_enabled       = false
}
```

Before we proceed to provision infrastructure using Terraform, we need to authenticate to Azure through the command line so that Terraform will be able to authenticate to Azure. One of the ways is through the Azure CLI. There are other ways to authenticate Terraform to Azure such as a service principal, or using a client secret or certificate; we will use the `az` command for simplicity purposes.

Authenticate using `az`:

```
$ az login
A web browser has been opened at https://login.microsoftonline.
com/organizations/oauth2/v2.0/authorize. Please continue the
login in the web browser. If no web browser is available or if
the web browser fails to open, use device code flow with `az
login --use-device-code`.
[
  {
    ...
    "user": {
      "name": "john@doe",
      "type": "user"
    ...
    }
```

```
    }
  ]
```

A browser link will open asking us to log in. Once done, we shall be able to execute commands using az but also, we should be able to run Terraform plans on our Azure account.

The commands to use are the same as we used in the previous chapter:

```
$ terraform init
$ terraform apply
```

The infrastructure will be provisioned and the Terraform state will be hosted on our local filesystem. However, there is a more managed way to store the Terraform state, distribute it to other team members, and prevent conflicts when accessing it.

Storing the Terraform state file

By just running the previous snippet, the state of the infrastructure provisioned will be kept locally. This is not something recommended in production, as the state should remain physically somewhere where it is feasible for multiple team members to access it. We shall use a storage account from Azure to achieve our goal.

Since we can create multiple resource groups, there will be a dedicated storage account for our Terraform plans.

Let's create the storage account through a script:

```bash
#!/bin/bash
RESOURCE_GROUP_NAME=guide-to-docker-compose-tf
STORAGE_ACCOUNT_NAME=guidetodockercomposetf
CONTAINER_NAME=tfstate
az group create --name $RESOURCE_GROUP_NAME --location
northeurope
az storage account create --resource-group $RESOURCE_GROUP_NAME
--name $STORAGE_ACCOUNT_NAME --sku Standard_LRS --encryption-
services blob
az storage container create --name $CONTAINER_NAME --account-
name $STORAGE_ACCOUNT_NAME
```

We can now change the provider so that the state will be stored on the storage account:

```
terraform {
  required_providers {
```

```
    azurerm = {
      source   = "hashicorp/azurerm"
      version = "=2.48.0"
    }
  }
    backend "azurerm" {
        resource_group_name   = "guide-to-docker-compose-tf"
        storage_account_name = "guidetodockercomposetf"
        container_name        = "tfstate"
        key                   = "terraform.tfstate"

    }
  }
provider "azurerm" {
  features {}
}
```

We set up the foundation to be able to use the ACI. We have a container registry available as well as a resource group where our ACI application will be provisioned. We will now proceed and adapt our application to make a deployment to ACI.

Deploying on ACI

Since we have a container registry in place, we can now upload the Docker images we shall build through our Compose application. The container endpoint can be retrieved by checking the registry on the Azure account.

The registry endpoint should be developerguidetocomposeecr.azurecr.io/developer-guide-to-compose.

We can now adapt our Compose file and set the name with the container images we shall push:

```
services:
  location-service:
...
    image: developerguidetocomposeacr.azurecr.io/developer-guide-to-compose:location-service_0.1
...
  event-service:
...
    image: developerguidetocomposeacr.azurecr.io/developer-
```

```
guide-to-compose:events-service_0.1

...

  task-manager:

...

    image: developerguidetocomposeacr.azurecr.io/developer-
guide-to-compose:task-manager_0.1
```

Changing the image to the full path of the Azure registry brings us closer to deployment; however, there are more adjustments to be made due to the nature of ACI. More specifically, there are limits on the ports that can be used on an application as well as the limitation in regards to the resources available.

One workaround we need has to do with ports.

As mentioned in the documentation (https://docs.microsoft.com/en-us/azure/ container-instances/container-instances-container-groups#networking), *"within a container group, container instances can reach each other via localhost on any port, even if those ports aren't exposed externally on the group's IP address or from the container."*

If Task Manager and the location service have the same port, then there is going to be a conflict and the following error will occur:

```
listen tcp :8080: bind: address already in use
```

To resolve this, we shall set Task Manager to use port 80:

```
  task-manager:
    container_name: task-manager

    ...

    image: developerguidetocomposeacr.azurecr.io/developer-
guide-to-compose:task-manager_0.1
    ports:
      - 80:80
    environment:

    ...

      - TASK_MANAGER_HOST=:80
```

By using this workaround, we shall avoid the port conflict.

Another issue also has to do with the readiness of some services. event-service will try to access the Redis service immediately after being initiated, and this might cause a failure. At the time of writing, depends_on is unsupported (https://docs.microsoft.com/en-us/ azure/app-service/configure-custom-container?pivots=container-linux#unsupported-options).

We can overcome this issue in a creative way. We shall adapt `event-service` and add some `sleep` before running the binary that is listening to the Redis stream:

```
event-service:
  container_name: event-service
  ...
  command:
    - /bin/sh
    - -c
    - |
      sleep 120
      /events_service
```

This is a simple workaround; however, there can be more effective workarounds that might even include a change in the code base and a health check adaption.

The next issue we should tackle is the resource limitations that come along with ACI.

In most regions, ACI comes with the limitation of four CPUs. By default, when a service is spun up, one CPU is allocated per service. This makes our deployment non-feasible, since we have more than four services.

In order to tackle this, we will adapt the resources used by the containers in our Compose application.

Docker Compose enables us to define resource constraints on the containers of a service. The available resource options are CPU and memory.

When we define the memory in the resources section such as `memory: 512M`, we instruct the container of the service to use up to 512 megabytes.

When we define the CPU on the resources section such as `cpu: 0.5`, we instruct the container of the service to use half of the available time of a processing core.

When defining the resources that a container will use, we have two options, `limits` and `reservations`.

The purpose of `limits` is to constrain the amount of resources to be used. This means that if our application has a burst of requests that require extra resources, the resources allocated will be capped by the limits specified.

By using `reservations`, we reserve the minimum resources that our application needs in order to be functional. If the resources are not available, our application is not able to operate.

For our application, due to the resource limitations, and because the default settings of our application will surpass the limitations (one CPU per service), we will proceed to use the resources configuration from Compose:

```
Setting cpu and memory limits:services:
  location-service:
   ...
    deploy:
      resources:
        limits:
          cpus: '0.5'
          memory: 512M
        reservations:
          cpus: '0.5'
          memory: 512M
  event-service:
   ...
    deploy:
      resources:
        limits:
          cpus: '0.5'
          memory: 1024M
        reservations:
          cpus: '0.5'
          memory: 1024M
  task-manager:
   ...
    deploy:
      resources:
        limits:
          cpus: '0.5'
          memory: 1024M
        reservations:
          cpus: '0.5'
          memory: 1024M
   ...
```

So now, we can push the images and execute requests toward our application. Before we push the Docker images, we need to authenticate to the recently created Azure registry.

First, we need to log in to Azure and then the registry:

```
$ az login
$ az acr login --name developerguidetocomposeacr
```

Now, it is possible to push to the container registries.

Let's build and push:

```
$ docker compose build --no-cache
$ docker compose push
...
Pushing location-service: cf8204bbc172 Pushing [============
====================================
==>]   557.8MB                                      265.5s
[+] Running 1/43ion-service: db60c013991c Pushed
                    265.5s
...
  Pushing event-service: 24302eb7d908 Pushed
                    269.7s
```

Pushing was a success; we now need to make it feasible for Docker to create resources on Azure:

```
$ docker login azure
```

This way, Docker will retrieve credentials in order to be able to interact with Azure. Onward, we shall create a Docker `context` that will interact with the Azure resource group we created previously:

```
$ docker context create aci guide-to-compose-azure
Using only available subscription : Pay-As-You-Go (c8ce802e-
f8d5-4634-b279-8d203a9c4882)
? Select a resource group  [Use arrows to move, type to filter]
  create a new resource group
> guidetodockercompose (eastus)
```

We are set up and can proceed to deploy our application to ACI:

```
$ docker --context=guide-to-compose-azure compose up
[+] Running 5/5
⊞ Group azure         Created                    10.1s
```

```
⊞ redis              Created                    144.7s
⊞ event-service      Created                    144.7s
⊞ task-manager       Created                    144.7s
⊞ location-service   Created                    144.7s
```

By using `docker compose ps`, we shall retrieve the IP for our application:

```
$ docker --context=guide-to-compose-azure compose ps
NAME                          COMMAND
SERVICE                STATUS                PORTS
azure_event-service           ""                    event-
service        Running
azure_location-service        ""                    location-
service    Running
azure_redis                   ""                    redis
Running
azure_task-manager            ""                    task-
manager        Running      20.237.67.76:80->80/tcp:80-
>80/TCP
```

By exposing a port through Compose, ACI will set up a load balancer and forward the requests from that IP to our service. Only one service can expose the specified port number.

As we can see, port 80 is exposed through the IP. We also configure the application to be accessible through a DNS domain:

```
services:
  task-manager:

    ...

    domainname: "developerguidetocompose"
```

By updating the application using `compose up`, we should be able to have the service exposed through DNS:

```
$ docker --context=guide-to-compose-azure compose up
...
$ docker --context=guide-to-compose-azure compose ps
...
azure_task-manager            ""                    task-
manager        Running               developerguidetocompose.
eastus.azurecontainer.io:80->80/tcp:80->80/TCP
```

We can now access our application using a DNS. Using our own DNS name is also feasible.

As we can see, we have made it and managed to access our application through DNS. We uploaded the containers to the Azure container registry, adapted our application, and made it feasible to be deployed to ACI.

Summary

In this chapter, we managed to deploy our Compose application to ACI. We created a Docker registry on Azure and pushed the Docker images to the registry. We then deployed our application to ACI without the need of provisioning any infrastructure. We did some essential modifications for our application to run and also tested our application through a DNS domain. So far, we have covered running Compose on two major cloud providers, AWS and Azure, while utilizing their container orchestration engine offerings.

In the next chapter, we shall try a different orchestration engine, the popular Kubernetes. We shall migrate an existing Compose application to a Kubernetes one.

13

Migrating to Kubernetes Configuration Using Compose

In the last few chapters, we deployed a Compose application to production using two cloud providers: AWS and Azure. We provisioned container registries and VPCs, as well as other cloud components such as load balancers. We created cloud-native applications and took advantage of the features provided by the cloud such as autoscaling, as well as load balancing.

In this chapter, we will focus on migrating our Compose application to a popular container orchestration engine: Kubernetes. Over the last few years, Kubernetes has grown a lot in popularity and it has a rich ecosystem of utilities and tools. There are many reasons nowadays for an engineer to choose Kubernetes as a container orchestration engine. This does not conflict with using Compose and Kubernetes. Compose can be the lightweight tool that can assist during local development, while Kubernetes can be the choice for production. In this chapter, we will make a transition from Compose to Kubernetes. We will identify the corresponding Kubernetes components to the Compose ones and we will generate the Kubernetes deployment configurations.

In this chapter, we will cover the following topics:

- Introduction to Kubernetes
- Kubernetes components and Compose
- Using Kompose to convert files
- Introduction to Minikube
- Deploying your application to Kubernetes

Technical requirements

The code for this book is hosted in this book's GitHub repository at `https://github.com/PacktPublishing/A-Developer-s-Essential-Guide-to-Docker-Compose`. If updates are made to the code, they will be reflected in this GitHub repository.

Introduction to Kubernetes

Kubernetes is a container orchestration engine. By using Kubernetes, we can automate the deployment, scaling, and management of container-based applications. Since it is open source, a Kubernetes cluster can be set up from scratch in your data center, whether it is an on-premise data center or in the cloud. Also, there is the option of using a managed Kubernetes cluster. Due to its growing popularity, every major cloud provider, such as AWS, Google Cloud, and Azure, has a managed Kubernetes offering.

When we deploy an application on Kubernetes, we deploy the application to an environment that has various capabilities that assist in provisioning, scaling, and monitoring that application.

Kubernetes provides a distributed and robust way to store application secrets and configurations. It offers a health check and readiness system for a deployed application. It can scale the application that's been provisioned as well as load balance the traffic. It also provides the necessary tools to monitor as well as debug an application. Finally, it provides a layer of service communication through naming a service discovery.

Considering these points, by deploying an application to Kubernetes, we expect the following to happen:

- The application is scheduled to be deployed and run on the node of the Kubernetes cluster.
- The application will use the configuration, secrets, and environment variables that are defined when deploying.
- Kubernetes will scale the application to the number of application instances we defined.
- Kubernetes will monitor the health and readiness of the application and will replace unresponsive containers.
- Kubernetes will load balance the traffic toward our application to the multiple instances that are running inside the cluster.
- Kubernetes will provide an extra layer of security by enabling and disabling traffic between applications.
- Kubernetes will provide a service discovery so that our applications will be able to seamlessly connect to other applications through the cluster without any manual intervention.

Now that we know more about Kubernetes and how this works for our applications, let's see how the Compose components of Kubernetes match with the ones we use for Compose and how we can make a migration feasible.

Kubernetes components and Compose

Our Compose applications are simplistic, but if we look carefully, they do have certain components in place. Those components have their corresponding Kubernetes components.

Compose applications versus namespaces

As we saw in *Chapter 11, Deploying Docker Compose to AWS*, an ECS cluster can host multiple Compose applications. Our Compose application, in a way, provides a way to group the resources that we provision on an ECS cluster. In Kubernetes, this is done through namespaces. Namespaces can help different applications share a cluster while being logically isolated from each other.

Compose services versus Kubernetes services

In the Compose specification, service represents the context of a service that is backed by one or more containers. As we know, when we define the service, we can configure the name of the underlying containers.

The equivalent of this on Kubernetes is the combination of Kubernetes Pods, Deployments, and Services. A Pod on Kubernetes is the smallest deployable unit of computing that can be deployed to Kubernetes. When deploying a Pod to Kubernetes, a singular container will be deployed to a Node in Kubernetes with the configuration specified. It is a single instance of the deployment we made, which leads us to Deployments.

Deployments on Kubernetes are a declarative way to define an application and specify the Pods that comprise that application, as well as the number of replicas that the Pod will have. Since we introduced the concept of replicas, load balancing is the next logical step to identify.

Services on Kubernetes provide an abstract way to access the Pods of a Deployment. In Compose, we can access a service directly by using the service name as a DNS. In Kubernetes, we need to define a Service that provides this abstraction inside the Kubernetes cluster. The Service will provide a single DNS name under which we shall be able to interact with the Pods of an application. Also, by using the Service, the traffic will be load balanced across the Pods of the Deployment.

Labels

In Compose, we can use labels – that is, key/value pairs that are attached to Compose components. Labels can be added to services, images, and every resource defined in the Compose application. With labels, we have a way to attach metadata to the Compose components. Kubernetes takes the same concept of labels and takes it one step further. In Kubernetes, labels are not limited only to some type of metadata information.

Labels can be attached to every component in Kubernetes. What makes Kubernetes labels important is their usage. Through labels, a Service can identify the Pods it should direct traffic to. Through labels, we can define rules that enable ingress and egress between Kubernetes Pods.

Compose networks versus network policies

In Compose, we have networks, which represent communication links between services. If we create a Compose application without a network defined, all the services will reside on the same network. This way, all the services can communicate with each other. If we define networks on Compose, for the services to communicate, they need to be on the same network. In Kubernetes, this is done with the help of `NetworkPolicy`. By defining a network policy, we can specify ingress and egress rules by using IP blocks, namespace matchers, or Pod selectors.

Labels have a key role in this since we can route traffic between Pods by using just labels. On the other hand, since we might want to be able to establish communication between multiple applications on a Kubernetes cluster located on different namespaces, traffic routing can also be defined using namespace selectors.

Now that we've provided an overview of the Compose components and their equivalent in Kubernetes, let's convert our existing Compose application into a Kubernetes one.

Using Kompose to convert files

Converting an existing Compose application into Kubernetes should not be difficult. A Compose application has a much simpler structure, whereas a Kubernetes deployment can get much more complex as Kubernetes comes with many features and capabilities.

As a use case, we will pick the application we built in *Chapter 5, Connecting Microservices*. We will just use the Compose file with some adaptations.

There is the option to convert a Compose application into the equivalent Kubernetes resources manually. Alternatively, a tool such as Kompose can be used. By using Kompose, we can convert our Compose application into the equivalent Kubernetes resources needed for our application.

Installing Kompose should be easy – we can just follow the instructions at `https://kompose.io/installation/`.

Before we jump into conversion, we need to make some adaptations to our application.

One of the things that Kompose needs is the version of our Compose file:

```
version: '3'
```

Another important part of the Compose service is to expose the ports. By exposing the ports, we make it feasible for Kompose to identify a port that needs to be exposed on a deployment.

In Compose, when exposing a port, it will translate into a public port so that it can be deployed to ECS and Azure.

In Kubernetes, exposing a port on a deployment serves documentation, just like EXPOSE on a Dockerfile. To expose a Pod in Kubernetes in a load-balanced form internally, a Service is needed.

Let's add the port on Redis:

```
services:
  redis:
    image: redis
    networks:
      - redis-network
    ports:
      - 6379: 6379
...
```

Now, let's add the port to the location service:

```
services:
  location-service:
    ...
    ports:
      - 8080:8080
    ...
```

Another adaptation we need to do has to do with health checks' start_period. In Kubernetes, there is the context of health checks but also the context of readiness prompts. Instead of waiting for when the health checks should start, a readiness prompt is provided to identify when the application is ready, and when the health checks should take effect.

Therefore, the applications with health checks should be adapted to this:

```
healthcheck:
  test: ["CMD", "curl", "-f", "http://localhost:8080/ping"]
  interval: 10s
  timeout: 5s
  retries: 5
  start_period: 5s
```

We are now ready to generate the files:

```
$ kompose convert
```

Provided we run on the same directory where the Compose file exists, we shall see the files being generated.

If we examine these files, we shall see three distinct types:

- Deployment
- Services
- Network policies

The content of our application and configuration resides in the deployment file.

If we inspect the `task-manager-deployment.yaml` file, we will see that the environment variables and their corresponding health checks are there. Also, take note that the port is exposed.

The next file to inspect is the service. Unlike Compose, where the application can be reached under the service name without any adaptation, in Kubernetes, we need to place a Service in front of it so that rooting traffic through a DNS name takes effect.

Then, we can see the network policies. Provided a network plugin is installed in Kubernetes, we allow traffic ingress from other Pods based on their labels.

In both Redis and the location network policies, the ingress between the Pods is routed by using the Pod labels.

Now, let's deploy the application to a Kubernetes cluster.

Introduction to Minikube

A way to run and test your Kubernetes deployments locally is through Minikube. Minikube is a local Kubernetes engine that we can deploy and test our Kubernetes application on.

Since Minikube is not a fully operational Kubernetes cluster, in the context of having some highly available master nodes and some node groups attached, we are limited to what we can test. For example, we cannot scale the underlying node group based on the increase of our workloads or spread the deployment of our applications to different availability zones. However, for our usage, it covers all the aspects of our application.

Based on your workstation instance, you can find the corresponding installation instructions (`https://minikube.sigs.k8s.io/docs/start/`).

If you have Minikube installed and want to start fresh, you can delete the previous container and start a new one:

```
$ minikube stop && minikube delete
```

A network policy requires a network plugin. This way, the rules will take effect and we can enable the service to communicate together or prevent them. By default, Minikube does not have a network plugin enabled. We shall use Calligo for our network policies to take effect.

Let's start Minikube:

```
$ minikube start --network-plugin=cni --cni=calico
```

If a failure occurs on macOS, you will see an error message similar to the following:

```
You might also need to resize the docker driver desktop.
```

In this case, resizing the disk image size of your Docker Desktop configuration can help:

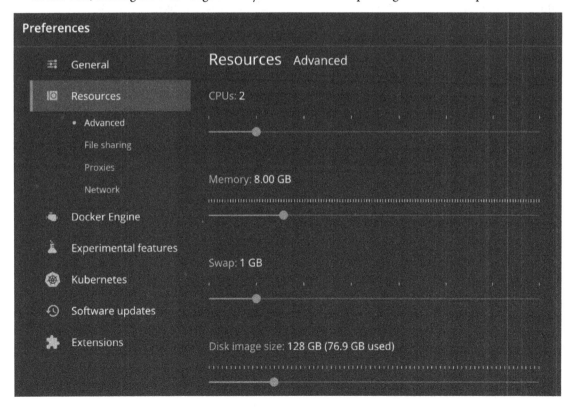

Figure 13.1 – Disk image resize

Let's test Minikube by applying a deployment:

```
$ kubectl create deployment nginx --image=nginx
deployment.apps/nginx created
$ kubectl get pod
NAME                            READY    STATUS     RESTARTS
AGE
nginx-8f458dc5b-z2ctk           1/1      Running    0
55s
```

Now, we can install the files that have been generated. The next step is to have Task Manager fully working on Kubernetes.

Deploying to Kubernetes

With this Kubernetes Deployment, we are closer to the goal of migrating our application. However, there is one thing we need to take care of and this has to do with pulling the images from a registry. As we have seen on ECS and ACI, it is essential to have a way for the container orchestration engine to be able to access the images from a registry. Since we are using Minikube, there is no need for us to provision a registry.

We can build our images and deploy them to the local Minikube registry. To achieve this, we shall point our build operations to the Minikube local registry.

We shall do this through the docker-env command of Minikube:

```
$ eval $(minikube docker-env)
```

Now, let's build and deploy those images toward that registry:

```
$ docker compose build
```

Applying the files we've generated previously should now be streamlined.

Let's start with the Redis deployments:

```
kubectl apply -f redis-deployment.yaml
kubectl apply -f redis-service.yaml
kubectl apply -f redis-network-networkpolicy.yaml
```

Since Redis is up and running, the next component we will cover is the Event Service. The only dependency the Event Service has is Redis:

```
kubectl apply -f event-service-deployment.yaml
```

The next Service that is also dependent on Redis is the location service:

```
kubectl apply -f location-service-deployment.yaml
kubectl apply -f location-service-service.yaml
kubectl apply -f location-network-networkpolicy.yaml
```

Last but not least, let's deploy Task Manager:

```
kubectl apply -f task-manager-deployment.yaml
kubectl apply -f task-manager-service.yaml
```

At this point, our application should have been fully deployed to Kubernetes.

Let's port-forward Task Manager:

```
$ kubectl port-forward svc/task-manager 8080:8080
```

Since our application is up and running, let's create a task by executing a request using `curl`:

```
$ curl --location --request POST '127.0.0.1:8080/task/' \
--header 'Content-Type: application/json' \
--data-raw '{
    "id": "8b171ce0-6f7b-4c22-aa6f-8b110c19f83a",
    "name": "A task",
    "description": "A task that need to be executed at the
timestamp specified",
    "timestamp": 1645275972000,
    "location": {
        "id": "1c2e2081-075d-443a-ac20-40bf3b320a6f",
        "name": "Liverpoll Street Station",
        "description": "Station for Tube and National Rail",
        "longitude": -0.081966,
        "latitude": 51.517336
    }
}'
{"created":true,"message":"Task Created Successfully","tas
k":{"id":"8b171ce0-6f7b-4c22-aa6f-8b110c19f83a","name":"A
task","description":"A task that need to be executed at the
timestamp specified","timestamp":1645275972000,"location":{"
id":"1c2e2081-075d-443a-ac20-40bf3b320a6f","name":"Liverpoll
Street Station","description":"Station for Tube and National
Rail","longitude":-0.081966,"latitude":51.517336}}}
```

We can check that the network policies work by tweaking the configurations.

Now, let's apply a `deny-all` network policy:

```
$ kubectl create -f - <<EOF
kind: NetworkPolicy
apiVersion: networking.k8s.io/v1
metadata:
  name: deny-all
spec:
  podSelector:
    matchLabels: {}
EOF
```

If we delete the Redis network policy, we won't be able to access the Redis database from the Pods anymore:

```
$ kubectl delete -f ./redis-network-networkpolicy.yaml
```

We did it! We just migrated our application to Kubernetes! The Compose services are running as deployments with a Kubernetes Service in front of them and we have our network policy rules taking effect.

Summary

In this chapter, we were introduced to Kubernetes, some of its basic components, and the corresponding Kubernetes components for our Compose application's components. Then, we migrated the Compose application to the necessary Kubernetes resources that are required to make our application operational. To do so, we used Kompose to streamline the generation of Kubernetes files. Then, we installed Minikube, including the network plugin Calico. By having Minikube up and running, we managed to deploy our application and test it.

So far, we have used Compose extensively for different occasions. We used it for day-to-day development, testing production deployments, and also to implement a Kubernetes deployment. At this point, you should be able to confidently use Compose for your day-to-day work productively.

I would like to thank you for choosing this book. Now, it is your choice where and how you will apply this recently acquired knowledge. Whether it will be your new cloud-native application, an environment for CI/CD tasks, or a local environment for your team, it is your choice.

That's it! This is the end of this book. Go ahead and build amazing things.

Index

D

`Packt.com`

Subscribe to our online digital library for full access to over 7,000 books and videos, as well as industry leading tools to help you plan your personal development and advance your career. For more information, please visit our website.

Why subscribe?

- Spend less time learning and more time coding with practical eBooks and Videos from over 4,000 industry professionals

- Improve your learning with Skill Plans built especially for you

- Get a free eBook or video every month

- Fully searchable for easy access to vital information

- Copy and paste, print, and bookmark content

Did you know that Packt offers eBook versions of every book published, with PDF and ePub files available? You can upgrade to the eBook version at `packt.com` and as a print book customer, you are entitled to a discount on the eBook copy. Get in touch with us at `customercare@packtpub.com` for more details.

At `www.packt.com`, you can also read a collection of free technical articles, sign up for a range of free newsletters, and receive exclusive discounts and offers on Packt books and eBooks.

Other Books You May Enjoy

If you enjoyed this book, you may be interested in these other books by Packt:

Mastering Docker - Fourth Edition

Russ McKendrick

ISBN: 9781839216572

- Get to grips with essential Docker components and concepts
- Discover the best ways to build, store, and distribute container images
- Understand how Docker can fit into your development workflow
- Secure your containers and files with Docker's security features
- Explore first-party and third-party cluster tools and plugins
- Launch and manage your Kubernetes clusters in major public clouds

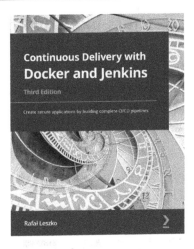

Continuous Delivery with Docker and Jenkins - Third Edition

Rafał Leszko

ISBN: 9781803237480

- Grasp Docker fundamentals and dockerize applications for the CD process
- Understand how to use Jenkins on-premises and in the cloud
- Scale a pool of Docker servers using Kubernetes
- Write acceptance tests using Cucumber
- Run tests in the Docker ecosystem using Jenkins
- Provision your servers and infrastructure using Ansible and Terraform
- Publish a built Docker image to a Docker registry
- Deploy cycles of Jenkins pipelines using community best practices

Packt is searching for authors like you

If you're interested in becoming an author for Packt, please visit `authors.packtpub.com` and apply today. We have worked with thousands of developers and tech professionals, just like you, to help them share their insight with the global tech community. You can make a general application, apply for a specific hot topic that we are recruiting an author for, or submit your own idea.

Share your thoughts

Now you've finished *A Developer's Essential Guide to Docker Compose*, we'd love to hear your thoughts! Scan the QR code below to go straight to the Amazon review page for this book and share your feedback or leave a review on the site that you purchased it from.

`https://packt.link/r/1803234369`

Your review is important to us and the tech community and will help us make sure we're delivering excellent quality content.

www.ingramcontent.com/pod-product-compliance
Lightning Source LLC
Chambersburg PA
CBHW060534060326
40690CB00017B/3491